Additional Praise for
Experiencing the American Dream

Experiencing the American Dream highlights that successful investing is about personal accountability and making a plan you can stick to over the long run. Investors can find peace of mind by rooting financial goals in their life aspirations. As a result, wealth planning becomes more tangible and investment outcomes are more closely focused and defined.

—Bob Merton
American Economist, Nobel Memorial Prize in Economic Sciences laureate, and professor at the MIT Sloan School of Management

Mark Matson understands investor psychology. He knows that, left to their own devices, many investors become their own worst enemies. They trade too often, cling to losers, and buy attention-grabbing stocks. They buy mutual funds they wish they had bought last year or, worse yet, stocks they wish they had bought yesterday. They get excited about the same stock at the same time which leads to losses more often than gains. Investors focus too little on fees and too much on recent market moves. And, without a clear plan, many investors panic when the market drops, as sooner or later it always does.

—Terrance Odean
Rudd Family Foundation Professor of Finance
Haas School of Business
University of California, Berkeley

I am much impressed with the contents of the binder you sent. Mark is certainly an outstanding exporter of my ideas I published.

—Harry Markowitz
The Father of Modern Portfolio Theory and Nobel laureate

Experiencing the American Dream is about what it really takes to be successful and what it means to live a life with purpose. Discover how to shed the old mindsets holding you back that lead to poor decision-making. Rediscover your values with money and what's most important in your life. What if money was used to make a difference for you and others?

—Chip Wilson
Founder of Lululemon

Making money isn't easy and investing isn't for the faint of heart. In a world where financial markets can seem complex, irrational and thoroughly unpredictable, *Experiencing the American Dream* offers sound practical advice to the nonprofessional investor. Grounded in Nobel Prize winning academic research and sound economic principles, this book empowers readers with a strategy to boldly take control of their financial future. Matson knows how to make money grow. He's achieved his American Dream and this book could help you achieve yours.

—Stephen Moore
Economist and former *Wall Street Journal* Senior Economics writer

When I was studying at the University of Chicago it became clear to me that the stock market is extremely efficient. When I first met Mark in 1991 he had the same epiphany. Matson has brilliantly communicated many of the critical modern financial theories in this book. Investors, who are drowning in a sea of misinformation, need this type of education now more than ever to make sound investing decisions.

—Rex Sinquefield
American businessman, investor, and philanthropist,
renowned as an "index-fund pioneer" for creating
the first passively managed index fund for the general public.
Co-founder of Dimensional Fund Advisors.

EXPERIENCING
THE
AMERICAN
DREAM

EXPERIENCING
THE
AMERICAN
DREAM

HOW TO INVEST YOUR TIME, ENERGY, AND MONEY TO CREATE AN EXTRAORDINARY LIFE

MARK MATSON

WILEY

Published by John Wiley & Sons, Inc., Hoboken, New Jersey.
Published simultaneously in Canada.

For general information on our other products and services or for technical support, please contact our Customer Care Department within the United States at (800) 762-2974, outside the United States at (317) 572-3993 or fax (317) 572-4002.

Wiley also publishes its books in a variety of electronic formats. Some content that appears in print may not be available in electronic formats. For more information about Wiley products, visit our web site at www.wiley.com.

Library of Congress Cataloging-in-Publication Data Is Available:

ISBN 9781394262045 (Cloth)
ISBN 9781394262052 (ePDF)
ISBN 9781394262069 (ePub)

Cover Design: Wiley
Cover Image: © Jane Kelly / Adobe Stock
Author Photo: Courtesy of the Author

SKY10076860_071724

Even before I knew you, I prayed for God to bring you into my life. Nothing that I have accomplished would have happened without you, including this book. Thank you for all of your love, support, courage, and wise counsel. This book is dedicated to you, my wife Melissa.

Contents

Author Note

In this book, I share with you my opinions and recollections. That means the ideas I present throughout this book are just one view. Others have different views, and I leave it to you to judge and decide which approach will best help you meet your goals. Don't take me at my word. I invite you to question, research, test, and challenge these ideas.

Also, you should remember that investing comes with risk, and every investor should think carefully about the risk they are prepared to take. Every investor should also know that past performance cannot tell you what future results will be. Even ideas and strategies that performed well previously have risk and can perform very differently in the future.

In this book, I have portrayed conversations involving others. I have put some of these in quotes to indicate that another person is speaking, but these reflect my recollection, not transcripts or recordings made at the time, and may differ from the recollection of others.

The individuals who have been kind enough to support this book by writing a foreword, epilogue, or other statement for the cover are not clients of my firm, Matson Money, Inc. None were paid or otherwise compensated for their supporting words. However, some of these individuals have a relationship with my firm. For example, Mr. Laffer sits on our Academic Advisory Board, and Mr. Lowe has previously appeared at Matson Money events. Others are associated with firms that provide investment products and services used by Matson Money. In some cases, there may have been compensation in connection with the relationship or prior events, or the commercial relationship between Matson Money and other firms. As a result, and because the individuals might receive some intangible benefits from being associated with this book (e.g., heightened brand awareness or reputational enhancement), there may have been incentives for their support.

This book contains data, graphs, charts, text, or other material reflecting the performance of a security, an index, an investment vehicle, or other instrument over time ("Performance Material"). Past performance, including any performance reflected in Performance Material, is not an indication of future results. This information is for educational purposes only and should not be used as investment advice. Investors cannot invest in a market index directly, and the performance of an index does not represent any actual transactions. The performance of an index does not include the deduction of various fees and expenses, or the impact of taxes, each of which would lower returns. Index performance results assume the reinvestment of dividends and capital gains. All investments involve the risk of loss, including the loss of principal. These risks may not always be mitigated through long-term investing or diversification. No investment strategy (including asset allocation and diversification strategies) can ensure peace of mind, guarantee profit, or protect against loss. PAST PERFORMANCE IS NO GUARANTEE OF FUTURE RESULTS. Additional disclosures and information are provided at the end of the book.

Acknowledgments

I have a love-hate relationship with writing. I love it when the words start to flow, and the copy seems to write itself. When there are stories that seem to come to life without restraint from deep within my subconscious. There are glorious times when this book seemed to effortlessly materialize, and it surprised me when it revealed itself. But there were also hard and lonely times of isolation and deep contemplation, and some days when it seemed I had nothing more to give. Hemingway, one of my favorite writers, reportedly said, "There is nothing to writing. All you do is sit down at a typewriter and bleed." There were many days when that is exactly how I felt.

I purchased an Air Steam trailer and parked it out in the desert and isolated myself to force the work into being. I knew, left to my own devices, there would be too many distractions anywhere else to get the job done. There were days I dreaded the 45-minute drive out into the Sonoran Reserve with the trailer ominously awaiting my arrival. And, yes, there were some particularly tough days recalling my childhood, or destructive screens I had in the past, that I would cry for 15 minutes before I could drag myself over to the computer. At times there were tears of joy at seeing how the American Dream had empowered me and my family. I experienced great happiness at sharing the mindsets and strategies that made it all possible. It was a cleansing process, cathartic and healing.

But although I felt alone in the trailer—I never really was. Every step along the way I had the support of family and friends who loved me and cared deeply about sharing this vision of the American Dream with others. I had colleagues and coworkers that had my back as I hammered away at the keys. No one writes a book alone, and I certainly did not write this one that way. It took a dedicated team to make it all come together.

First, I would like to thank my wife, Melissa, whose strength, courage, and endless support pulled me through. She kept things together at the office, the house, and for our kids and family. When I was at my worst, she was at her best. She pulled us all through it and she did so with grace and kindness. Melissa and I have eight children, and they all supported and encouraged me through the writing process: Mallory, Alex, Amanda, Jordyn, Jonathan, and Madison, and two of whom are still at home, Gage and Liam. They never complained when I had to work late, or on the weekends, or when we had to cut a family vacation short to meet a writing deadline. I thought about you all often when writing about freedom, fulfillment, and love in families. Being your father is a great gift and privilege.

To my mother and father for making the American Dream real for me. I am thankful for all your help in sharing family history with me. The love and care you gave me and my brothers, growing up and today, will always be remembered and cherished. I have deep appreciation for the hard work and stern determination of my grandparents and ancestors who sacrificed in the coal mines and chemical factories to make this future possible.

To my good friend Rob Lowe, whose honest review of the original introduction sent this book in a new direction. It was on a cold and dreary night in Cincinnati, while visiting my mother who was in the hospital, that you implored me to trash the original beginning and share the story of my family saga based on the conflict between my father and grandfather, what it began to reveal to me, and to keep that story line going throughout the work. It was a bold suggestion, and it took courage to share it with me. The book took on a new dimension in that moment—it became a form of memoir, self-development, and investing book. When you read the new version, I asked you to write the foreword, and you graciously accepted. I am deeply grateful for your direction and support—two Ohio boys working together to share the American Dream.

The American Dream is based to a large part on the ideals of capitalism and free markets. Many years ago, my father and I were invited to hear one of its greatest defenders speak at an investment conference. When we first met Dr. Art Laffer, inventor of the Laffer curve and recipient of the Presidential Medal of Freedom, we formed a close and dear friendship. Dr. Laffer has been on the Matson Money Academic Board for many years. I could think of no one better to write the epilogue and share how this great country was united in the past under President Reagan and how with determination and strong leadership it could be again in the future. I am profoundly thankful for your contribution to this work and your lifelong friendship.

A special thanks to Ryan Dempsey, my writing coach and collaborative author; Kevin Anderson for all your dedication and support; and Roger Scholl for your wise editing and structure additions. To my book agents at Book Highlight, Mat Miller, Peter Knox, and Brian Morrison, thank you for catching the vision and running with it before anyone else did. I am deeply grateful to Wiley Publishing: Kevin Harreld for driving and managing the process and Vithusha Rameshan, Susan Cerra, Premkumar Narayanan, and Susan Geraghty for your careful and diligent review and editing.

To Dr. David Eagleman for your groundbreaking research in neuroscience and Terreance Odean for all your input on behavioral biases and blind spots investors face. I am grateful for the groundbreaking and Nobel Prize–winning research from Eugene Fama and Dr. Harry Markowitz, who served on the Matson Money Academic Board. To my longtime friend Dr. Lyman Ott for your statistical research into portfolio design and investment premiums. Mark Fortier and his whole team at Fortier Public Relations for getting the message out to the world; you guys rock.

I will be forever grateful to my colleague and early mentor in the money management field, Dan Wheeler. Dan died two years ago but will always be remembered as an early innovator and truth teller in this industry. He largely revolutionized how advisors see investing. In those early days Dan fought in the trenches with me to spread the academic investing principles in this book.

A special thank-you to the visionaries at Dimensional Fund Advisors: without your research and support this book would not have been possible—and to Rex Sinquefield, who first taught me the message of free and efficient markets. With deep gratitude for David Booth and his leadership, innovation, and drive. David Butler, Bryce Skaff, and Hunt Carnes, thank you for your participation in the American Dream Summit and your incredible support in reading, sharing insights, and constant support to make this project come to life. Savina Rizova, thank you for your participation on the Matson Money Academic Board and your excellent research into equity markets and portfolio design.

To Michael Lane, thank you for years of friendship and for your time and contribution at the American Dream Summit. Your dedication to Matson Money is greatly valued.

Any CEO who hopes to write a best-selling book is faced with a monumental question. How do I get the time, energy, and resources to write a book, and at the same time get my company to continue growing while I am in seclusion? Not an easy problem to solve. I am fortunate because I am blessed with an incredible team at Matson Money. Not only did they "keep the doors open," they also grew the business while I penned this book. Josh Crawford, thank you for your bold and decisive leadership. Your commitment and dedication to the mission was nothing short of extraordinary. I appreciate your insights into the American Dream Experience and your help making this a better book. While I wrote, you coached, trained, and developed the advisors and internal staff, and then when it was time to release this book into the world you drove the promotion and sales like a true pro—bravo. Thank you for your undying dedication and support. In addition, I'm grateful for the hard work and commitment of all of our internal trainers: Luke Rowand, Cody Zindler, Matt Der, and Chris French.

To Dan List, Matson Money president, thank you for your specialized abilities to guide and manage this company as well as your keen eye for editing and all the technical aspects of this book. Without your 30+ years of dedication this book could never have happened. Matt Matson, Alex Crawford, and Michelle Matson, your creativity, loyalty, and hard work leading the teams that are crucial to drive technology, inspire the brand, and develop world-class training and development materials is unsurpassed.

Joel Autenrieb, Heather Nelson, Jeb Snyder, and Rick Wiehe, I am grateful for the trust you all inspire in directing operations and the management of over $10 billion for investors all over North America. It is a monumental responsibility, one I did not have to worry about as I created. And to the whole Matson Money staff, you are an amazing group of people dedicated to saving investors from speculating and gambling with their money and restoring the American Dream. Your efforts are nothing short of heroic.

And finally, to all the advisors and coaches who have fearlessly dedicated themselves to reviving the American Dream and standing up to the Wall Street bullies. Matson Money would not exist without you and the powerful relationships you form with investors. Investors should be grateful to have your bold stand to eliminate speculating and gambling with their investments and helping them focus on their true purpose for money and life.

Although the list is too long to mention each and every advisor by name, I want all of them to know how thankful and grateful I am for all your hard work and dedication. Some of the names are Greg Heaton, Paul Winkler, Scott Iles, Greg Hammond, Gretchen Stanger, Jan Hundrieser, Steve Rice, Mark Connely, Dale Pieper, Ira Work, Leonard Raskin, Evan Vanderway, Steven Vanderway, Ron Geer, Randall Davey, Joe DeLisi, Greg Black, Eric Fischesser, Jonathan Walker, Evan Barnard, Dan Hill, Justin Makris, Jerry Lyke, Roy McBryar, Becky Walker, Eric Rothman, Jeff Furest, Nick Sklenar, Bryan Weiss, Anthony Bonanno, Evan Barnard, James DesRocher, Robert Ramsey, Doug Guernsey, Ben Harper, Cutis Erikson, Eddy Shum, Michael Sarcheck, John Crane, Jim Wood, Maria Kuitula, Phyllis Wordhouse, Charles Stegall, Michael Evans, Leo McGrath, Chris Strehle, Charles Petit, Sophal Petit, Vincent D'Addona, Justin Perdomo, Margaret Whittkopp, Joe Matson, Fred Taylor, Darren Violette, Rhen Stevens, John Bellino, David Rosen, Doug Brauer, Michael Babin, David Shapiro, Jeffrey Mathies, Kristin Niedermeyer, Omar Pereira, Daniel Salazar, Darrell Doi, Eric Harris, Garry

Liday, Jason La Vigne, Jeffry Weldon, Carolyn Weldon, Larry Stone, Phil Webb, Richard Cochran, Kevin Kinghorn, Thomas Blottenberger, Vincent Del Franco, Jamie Smith, Bridget Mackay, Joshua Thomson, Misha Schryer, Raymond Martin, Stephen Stark, Steven Smartt, Alexander Cohen, Angela Ashley, Alex Ash, Anne Sawasky, Ben Harper, Brandt Jordan, Christopher Smith, David Yaw, Harry Shepler, Jacqueline French, Jay Bartley, Jeffrey Field, Jim Wood, John Weyhgandt, Larry Mott, Mat Miller, Michael Ringel, Michael Sailor, Saul Cohen, and Walter Lynn.

Foreword

Rob Lowe

Today, many are filled with questions that would've seemed out of place only a few years ago. Is there a better future? Is the system rigged? Is the American Dream dead? With this book, my friend Mark Matson has written an eloquent answer to these and other pivotal questions of our time. A natural storyteller, Mark will take you on his journey from a child one generation away from the most brutal poverty imaginable to becoming a leading thinker in finance and practitioner of investments at the highest level.

Like me, he was a Midwestern boy who somehow knew there was more to life than what he saw around him. And, like me, he didn't really have extended family, friends, or mentors to guide him on his quest for answers, so he went to work to find the answers himself.

Regardless of what you may initially think, we all have a complicated relationship with money. We love it. We hate it. We want it. We're scared of it. We're scared of not having it. We're confused by it. Some are "ambivalent" about all things financial, which as you will see, can be the most dangerous mindset of all! Mark has spent a lifetime exploring all these matters. And in the end, it boils down to this: How do you make money, and for what purpose? These are simple questions with complicated answers that are found on a journey that is surprising, entertaining, educational, and sometimes emotional.

Mark once asked me to tell him my first memory about money. He explained that those early experiences often set the stage for a lifetime of patterns and prejudices.

I thought of being seven years old and seeing bills and receipts spread across our dining room table for weeks on end. I remembered my stepfather up late at night poring over them doing God knows what. I had a vague notion that this unending and stressful chore was what was required to "balance a checkbook." And I remember thinking, "I want no part of that." How could anyone possibly crack this clearly complicated, mysterious, and emotionally exhausting code? I already hated math, so without knowing it, I setup a bias against and an avoidance of personal finance.

Mark then asked me about the first time I made any money. I thought of my paper route. I must've been 11 or 12. I loved getting up early and tossing the paper onto the neighborhood porches. I had a number of techniques for maximum accuracy and a soft landing so the paper would stay folded and not blow away. I had pride in my work. But getting paid was another story. My route happened to be in a part of town that was, shall we say, "in transition." Timely payment of the paperboy was not really a thing. It was next to impossible to collect, and so I never came close to making the money I should've. In fact, my father, who was a lawyer, decided to teach me about the legal system by filing a small claims case on my behalf. We showed up to court together, both of us in suits. My customers (the defendants) did not. The judge ruled in my favor. And I still didn't get paid. My takeaway was this: hard work gives a sense of fulfillment and purpose, and it can be fun. But on some level, unlike the other paperboys who did get paid, maybe I wasn't worthy of being compensated. Years later, when every other actor on *The West Wing* received a raise but me, I realized I was right back on my paper route.

As you read on, you will learn how to identify your own monetary worldviews and how they affect your life. You will learn incredible tools to navigate the road to financial well-being. You will see through a new lens the value of capitalism, the true nature of the public markets, the agenda of the financial media and the business of financial planning and investment. But along the way, you will also discover even larger truths and tools, and you will find the answer to the question, What exactly is your American Dream, and how do you manifest it and use it to power your true life purpose?

Preface: Whispers of the American Dream

"You have nothing. Great, start from nothing."

—*Hiral Nagda*

The American Dream saved my father, but it couldn't and wouldn't save my grandfather.

The term *hillbilly* was first formally used in *The Railroad Trainmen's Journal* IX (July 1892). It was meant to signify people who dwell in rural mountains in areas of the United States. In 1920, the *New York Journal* described them as people who live in the hills, have no means to speak of, dress as they can, and talk as they please. It was also inferred that they often shot their guns in the air for no apparent reason. Outsiders 60 years ago might have considered them to be backward, unsophisticated, or even ignorant.

This was the clan I was born into on August 23, 1963, in the city of Charleston, West Virginia. My parents were young with few prospects to create anything like real wealth to escape the destitute poverty they were born into. They worked tirelessly there until I was five years old, and then took what was then known as the *hillbilly highway* leading into the heart of the Midwest. Many others migrating out of the hills of West Virginia and Kentucky would make it as far as Chicago, Indianapolis, or even Kansas City. My parents stopped and laid down roots in Greenhills, a suburb of Cincinnati, Ohio.

It was here they would take a stand to build the American Dream for me and my two brothers, Brian and Matt. My father took a job selling insurance, and he was a natural at it. He was outgoing and friendly as well as intelligent and dedicated to learning his chosen field. Joe was a firm believer in the benefits of education and knowledge, something he would pass on to his children. As my parents committed to raising their family out of poverty with the hopes of reaching the lower middle class, we visited West Virginia every summer to see our relatives.

My father would get us up early on a Saturday morning, rousing us from our sleep well before the sun came up. Any kind of vacation was a big deal. We eagerly piled into the family station wagon. My father drove, and my mother rode shotgun to navigate. Watching them hold hands through most of the drive made me feel safe and protected, as if the love of our family was unshakable and forever enduring. It felt like there was a protective force field around our family.

Because it was long before cars beeped to remind passengers to put on seatbelts, my brothers and I grabbed our sleeping bags and jumped through the open hatch at the back of the car. We settled in for the three-and-a-half-hour ride with no concern for our safety—oblivious to the very real danger of traveling without seatbelts in the back. As kids, we thought we were indestructible.

My brothers and I constantly fought verbally and physically; most of the time it was playful, but sometimes it turned angry and aggressive. We were testing our mettle on each other. Steel sharpens steel, and we were making each other stronger.

Once we set out, my brothers and I tried to go back to sleep but failed miserably; we were far too amped about the trip. Mom was in control of the radio and, for now, had it tuned to Q102, the light rock station. Jimmy Buffett had written a killer song that summer that spanned both country and rock—"Margaritaville." It seemed to be on a constant loop on our station, and we belted it out every time it came on. We couldn't get enough of it. We had an unspoken contest to see who could sing it the loudest, and it cracked us up every time. My dad could only take so much of this. My mom was more tolerant, but only a little. She was still laughing with us when dad yelled, "Stop that 'grab assin' around. We didn't raise you all in a barn. If I have to stop this car, you are going to get it."

Out of all the times he threatened to stop, he never did. But we still took him seriously. My dad usually meant business if he had to raise his voice. None of us wanted to test his resolve. We talked quietly in the back, watching the darkness go by outside the windows. Then, throwing caution to the wind, we raised the volume back up when good old Jimmy came back on. We loved that Caribbean groove. Instead of Charleston, West Virginia, we imagined we were on our way to Key West.

I always admired the way Jimmy ended the song, claiming his fortunes were his own damn fault. He didn't buy into the narrative of blaming someone else for his trouble or woes. He refused to take the easy way out or thrust his failures onto anyone else. The "wasting away" was on him. And if it was going to get fixed, well, that was going to be on him, too. He went from it's nobody's fault, to it could be his fault, to it is his fault. He moved from being a helpless victim of mysterious forces to someone who discovered they have a say in life.

Driving east, we watched the sun rise through the front windshield. An hour and a half into our trip, we found ourselves in the middle of Ohio farm country, surrounded by chest-high cornfields. It would be years before cell phones were invented. We were so cut off from society that it felt like we were on the dark side of the moon. As time dragged on, we became restless.

Tired of lying down in the back of the station wagon, Brian, Matt, and I climbed over the back seat to sit in the middle of the car. Now, we were directly behind my parents. Brian was 11 and Matt was 8. Matt and I made sure Brian was in between both of us. Bored, Brian pulled out a book from his Greenhills' Pioneers gym bag and began to read. If you tried this on the roller coaster roads of West Virginia, you were asking to be car sick and puke, but the highway was relatively straight as the twisted roads of the mountains were still an hour off. The minute he started reading *The Lord of the Rings*, that was our cue to begin our chant. "Mr. Reader, Mr. Reader, Brian is a Mr. Reader." Matt and I thought this was gold. Of course, it was juvenile and immature, but we got a big kick out of it. On occasion, Mom had let us stay up late to watch *Saturday Night Live.*

And in our minds, we were up there with Chevy Chase and Richard Pryor, my favorite comedian. The Cone Heads and Blues Brothers had nothing on us, or so we thought.

If the chant alone didn't work its magic and cause Brian to yell at us and then (despite his resolve to stay upset) give into the odd and ridiculous nature of the moment before finally laughing with us, we took it to the next level. Both of us scooted as close to him as we could, sandwiching him between us, and began to flex our butt cheeks in rhythm with the chanting. Brian could have quite the temper, and this usually got him going. This move was sure to make him lose his concentration on Frodo of the Shire.

"Stop it! You guys are annoying," Brian threatened.

"What's that matter, Mr. Reader," Matt goaded him.

"Mom, Dad they are bothering me. Make 'em leave me alone!" Now he was bringing in the big guns.

My dad said, "Leave him alone, or you are going to get it."

"You are going to get it" was his mantra along with "Don't worry about what anyone else thinks." We must have heard each of those 10 times a day. If he stopped yelling, we knew we had a real problem, but my father's voice had slightly cracked, so I could tell he was trying to hold back a laugh. We were a very loud family, except for my mother. She was our calm sense of reason when we everyone else got carried away.

True to script Brian finally relented and began to laugh hysterically. The three of us, together with my father, guffawed good heartedly. Mom sat in the front seat smiling.

My parents had not only gotten us out of the hollers of West Virginia, they had created a home where it was okay to have fun and even be silly. We were encouraged to express ourselves and talk openly about almost anything. We didn't have a lot of money, but the sense in our family was that life could be an adventure and should be lived fully. Our home had an optimistic energy surrounding it. The type of energy that said anything is possible if you are willing to work hard enough. It was the idea that things could and would always be better. My parents communicated to us that life was full of wonder and something to be celebrated and cherished.

It was meant to be lived fully and boldly, on your own terms. Let others be damned if they didn't approve of or understand us.

We saw other people as basically good. There was a sense that the future would be better than the past. But the past was also full of stories, and if you kept those stories close to you and alive, you could glean messages and powerful meanings from them. If you cared to pay attention, there were morals and meanings to stories that could make your life and the lives of others better and fuller.

When the empowering moral of a story was not self-evident, my father taught us that we needed to dissect and tease it apart to find the brass ring of knowledge concealed inside. He also cautioned against letting the pursuit of "the ring of wisdom" you were seeking (like Gollum and his obsession with his *precious)* overtake and dominate your life. In other words, never forget your roots or "get too big for your britches" and let money, power, or fame destroy you. He often chided, "Never take yourself too seriously. You must be able to laugh at yourself."

My parents lived by example, and although my mom was often quiet and reserved, my father never met a person he didn't immediately make his friend. He was a masterful storyteller. When he got going, he would tell one story after another, the flow of tales creating a kind of magic all their own.

There was a sense of playfulness in our family, and although my father could be a firm disciplinarian, there was constant humor and lightheartedness that pervaded everything, including long car rides. Still, a somber feeling penetrated our station wagon as it barreled to the West Virginia border. I made it my job to listen to everything my parents said about life. I was already working on my grand opus on how life works and its many hidden mysteries. I sat calmly as my parents discussed the health and welfare of our family back in the hills. Who had gotten ill, lost a job, or retired? What were my grandparents doing? How did they live their lives? Had anything changed since our last summer trip there? It seemed that someone was always moving back up in the hollers after trying a brief stint in the city. There was much talk of "the family" avoiding any real change or new ways of thinking. My parents called it being "stuck in their ways." In retrospect, that was a kind way of putting it.

The West Virginia countryside was beautiful—the mountains a stunning shade of green with outcroppings of stone atop them. The valleys were adorned with crystal blue rivers and streams. The sight of them always prompted me to say, "I bet there is a ton of fish in there!" My dad would retort, "Bet your ass there is. I used to fish that river when I was your age." But against this idyllic canvas of beauty, the poverty was soul crushing. As we drew closer to my grandfather's house on my dad's side, the more desperation and decay appeared.

Many of the creeks (pronounced *cricks*) that ran up the hollers were full of discarded trash. Among the scenic boulders were rusted out truck beds, discarded washing machines, moss-covered couches, and broken-down furniture of every kind. Signs saying NO DUMPING were routinely ignored. Some were pockmarked with bullet holes. Deeper into the hills, many of the houses were little more than rundown shacks, mixed in among the trailers.

Many of the yards were strewn with garbage. Front porches were constructed with plywood with crumbling cement steps leading up to them. It was an army of broken TVs, discolored wash tubs, rusting bikes, and plastic toys strewn here and there. The sides of these houses were virtual junkyards, covered with grime and neglect—lawnmowers, plows, cars, bed springs, shovels, broken satellite dishes. These were not relics some industrious mechanic was ever going to repair or make useful again, nor were they antiques to be sold off at a local flea market. This was dirt and filth. It was clear the owners had no intention of cleaning up their yards. Neglect had turned into a tapestry of despair.

It was a direct affront to the ideals and principles my father and mother were working so hard to build into me and my brothers. It shocked me to see children playing among the refuse. One of my mother's favorite sayings was, "It is not a sin to be poor, but it is a sin to be dirty." My mind raced with questions. *Why did these parents make their children live this way? Why couldn't they carry their debris past the front porch?* And the overarching question: *How can people live like this?* The poverty I could understand. But the lethargy and the sloth were unfathomable. If I could unravel this mystery, maybe it could shed light on how to avoid a similar fate, or perhaps, and this seemed like a long shot, how to live an empowered life of greatness and abundance. Clearly there was a sort of learned helplessness hiding in plain sight here, some kind of failure to be and act. There was a debilitating lack of freedom in evidence. My fellow hill people were not being held physical prisoners, but there was something destroying their lives and keeping them firmly in place.

My grandmother was a large woman; she only wore overflowing V-shaped mumu dresses. My grandfather was, by contrast, "thin as a rail" in West Virginia speak. Later in life, Retha would contract throat cancer from smoking and subsequently had her larynx removed. In its place, the doctors created a hole where she could breathe. She talked with the aid of a vibrating handheld device she placed on her neck, below her jawline. Her artificial voice sounded robotic and harsh, and it was hard to understand what she was trying to say. The hole and the machine voice made me afraid of her. I felt terribly guilty for those feelings. But when she hugged me in that mumu dress with her big arms, that feeling of fear disappeared and was replaced with a feeling of safety and warmth.

I was not close to my grandfather—I'm not sure that anyone was. My father told me his father never hugged him or told him he loved him when he was a boy. He showed a similar disinterest in his grandsons. Sitting on his lap, getting "horsy rides" and being told he was proud of us and loved us was not in his range of affection or emotion.

As we approached Grandpa George's house, a strange silence crept over our family. There was no Jimmy Buffett on the radio, and no teasing and taunting of each other in the back seats. My mom and dad were equally still in the front seat. There was no sense of joy or happiness to finally be here. Our presence there felt more like an obligation than a celebration of family. I felt the silence pressing in on me.

The small yard was surrounded by a chain link fence that resembled a prison fence. It was unnaturally high, well over my head. It gave the impression that it was meant to keep something or someone in, not just out. Inside the fence, the yard was full of overgrown weeds. No one had planted flowers or decorative shrubs of any kind. One struggling maple in the front yard looked diseased and spindly, barely hanging on. Between the weeds were patches of dirt and gravel. Not even the dandelions and witch grass could find purchase there.

In the back yard, there was an upside-down metal flat-bottom boat resting on two dilapidated sawhorses that were devoid of paint and had rusted nails sticking out in all directions. Whoever built them never heard my mom's speech about the dangers of tetanus, or just didn't care. That boat and sawhorses were what my dad called "an emergency room trip waiting to happen." He wasn't wrong. However, that didn't stop us from playing on it.

The boat was the one thing in the yard that signified any real life or sense of fun and adventure. It must have been made of galvanized steel because it had managed to escape the rusty destruction all around it. But it had been abandoned for years. We wondered when the boat had last been in the water and seriously doubted it would ever be there again.

We arrived just in time for breakfast: fried eggs, sunny-side up, bacon, and strawberry jam. Grandma also served up homemade biscuits and sausage gravy. The smell of the food cooking mixed with the stale odor of Marlboro Reds. My wife's family is from the hills of Kentucky; she also remembers growing up with the taste of biscuits and gravy for breakfast. When we go out to brunch, we often order biscuits and gravy with the hope of reliving the nostalgia of our childhoods. Sometimes there is just no going back, even when you want to.

Walking in the front door, the home smelled as musty and decayed as a crypt. The walls were a flat yellow from years of cigarette smoke. The carpet was a faded tan. There was a couch and two chairs in the living room—a short hall with a bathroom and two small bedrooms on either side.

My brothers and I found an old wooden wardrobe in our assigned bedroom. We imagined we were entering the land of

Narnia when we climbed inside and closed the door behind us. We found an old flashlight and used it to go through boxes of black and white pictures of people we didn't recognize. Everything had the pungent odor of mothballs. Most of the pictures were of people standing outside of their shacks up in the hollers, or coal miners covered in dust, wearing hard hats with headlights. They were always standing by a mineshaft. And no one ever smiled. They clearly lived a hard life, and they were all deadly serious.

There was no central air in the house, but we did benefit from a large window fan as we slept that night. Its steady hum soothed my ragged emotions. The fragrance of trees and grass filled the room. But there was also a faint scent underneath that I couldn't quite place, but it reeked of death.

Everything about their home felt gray and subdued. Even the television in the living room sat ominously still. I never saw it on. Its giant ashen eye looked unblinkingly into the room, like a monolith. To this day, I do not know if it was broken. It sat unmoving like a giant lifeless cyclops. The home was totally unremarkable with one exception. On top of the television was perched a beautiful gold skeleton clock, with an engraving on the name plate:

George S. Matson
In recognition of 25 years of
Loyal Service
Union Carbide Corporation 1970

It was an Atmos clock, designed and made by Jaeger LeCoultre, a marvel of human engineering and art. It never needed winding, batteries, or electricity. Instead, it functioned by harnessing different air temperatures and pressures in the room. Ethyl chloride in the clockworks expanded and contracted, causing the bellows to fill up with compressed air, which in turn drove the mainspring. Adding to the wonderment of the clock, its internal mechanisms were visible from every angle, encased in glass. Somehow, the inventors had created the illusion that the working parts were floating inside the glass.

It was a stark reminder of the possible miracles of the world outside of these walls. It was the only object in the house that showed any signs of beauty or hope. Sadly, it seemed my grandparents were content to waste away in their self-imposed prison until the bitter end. Retha died some years later, and George soon followed. When their children inventoried the contents of the meager home, the clock still sat on the television, stubbornly ticking, untouched by the ravages of time. In a world of poverty and what Thoreau called "quiet desperation," the clock shined on, its gold magnificence seemingly mocking the spent lives of the house's former inhabitants.

After four days visiting our grandparents, aunts, uncles, and cousins in and around Charleston, we were ready to head home. The longer we stayed, the more our country twangs reasserted themselves.

Our mood was somber as we made our way home to the new world we were creating together. But we were grateful to escape. The poverty I had experienced in the Appalachian region was persistent and generational, like a curse, embedded in the very fabric of the land. Even as a kid, I felt it was not just a function of a poor economy, but something more sinister and permanent hidden under the surface.

Despite the poverty, my family who remained there were friendly, generous, and kind. As my mom would say, "They don't have much, but they would give you the shirt off of their back if you asked them for it." But what I remember most about them was that they were incredibly funny. They were always joking and laughing. Most of the humor was improvised while some of the jokes were repeated year after year, like a vinyl record player stuck in the same groove, repeating the same phrase of music, over and over. But the humor was part of who they were. I suspect they used it to cover the harsh reality they faced, like a salve to heal their wounds. Perhaps it was their way of pretending things were not really "that bad after all."

Was there something about the way my hillbilly brethren thought about themselves, the world and money that locked them into the reality they experienced? If that was true, then how had my parents escaped the same fate? Matt was more interested in playing a Mattel electronic football game, and Brian wanted to read about Frodo as he approached Mordor, but I was fascinated to learn more, so I peppered my parents with questions about their childhood.

"Dad, what was it like when you were a kid?"

"Well, I was born in 1940 just after World War II started. One of the things I first remember was hiding under the desk at school when they practiced air raid drills. Like that was going to work."

"What was your first house like?"

"When I was four until middle school, we lived down by the railroad tracks. You think grandpa's house is small, you wouldn't believe how little this thing was. It only had one real bedroom, a short hall, tiny kitchen and one room in the front of the house. It was me and my two brothers, Virgil and Don, my two sisters, Evelyn and Mary Ann. The boys shared one bed and my sisters another. Then there was my mom and dad as well as her parents. It was super tight."

"What did you do for heat in the winter?"

"We got our heat from a little gas fireplace that was lined with asbestos."

I found out that this was quite common in the early 1940s. The fireplace was made from a material called Transite; it was a form of cement mixed with asbestos filaments. This was used to insulate the fireplace walls from catching fire and push the heat out into the room. Like so many things in Charleston it was extremely dangerous. Over years of use, the cement becomes friable, so the asbestos strands heat and becomes airborne, causing lung and other cancers.

"The back of the fireplace burned red hot in the winter, and as the temperature dropped, all of us huddled around it. Damn, it got cold."

He paused, deep in thought, before continuing. "Our front porch dumped into our front yard, which was gravel and barely big enough to hold my dad's beat-up car. George worked odd jobs to support us the best he could. For a while he drove a Tillings ice cream truck. He sold Johnson floor wax door to door and then offered to buff his customer's floors if they bought from him. He even tried his hand at photography. He had to drop out of school when he was in fifth grade and work in the coal mines. But nothing seemed to stick."

"What did the rest of the house look like?"

"The dirty wallpaper was peeling off the walls. There was no drywall, only wooden planks. There were some cracks in the walls you could practically see through, and the wind whipped through them in the winter. The floors were partially covered in linoleum. On the areas where there was heavy foot traffic only tar paper remained. Our back door was only yards away from the train tracks. There were no passenger trains. Most of the trains hauled coal, but the hobos who rode the rails would knock on our back door and ask for work. They meant it, too. They weren't just looking for a handout. We never had work for them. George could barely find work for himself. But my dad always gave them a can of beans or corn—anything we had on hand, which usually wasn't much. He was always generous that way.

"In the fall before school started, we each got a new pair of shoes. We wore them all winter. But we only had one pair, so if we had grown when summer rolled around, they were too tight, and we would go barefoot. If we couldn't scrounge up the money to get a pair at Goodwill, we just went without. Dad wasn't much for charity. But if things were really tight, we might get some clothes from the Salvation Army. If we didn't have any money for Christmas presents, they would give us a little bit of fruit and a small toy, maybe a yo-yo or marbles." Both of which my father learned to master.

My mother added, "We were dirt-floor poor, and cane-switch raised. Many times, we didn't have enough to eat. We got commodity cheese from the government. Most nights, my mom cooked up a mess of white beans. If things were going well, we would have cornbread with 'em. Your aunt Linda and I liked to slather butter on the bread then crumble it up in our beans. We had meat once a week on Sundays, usually chicken. If we didn't have white beans, then we got potatoes. In the summer, we had a small garden, but it's hard to farm on the side of a mountain. My mom would have us girls string green beans; they were the best if we had a ham bone to cook up with them. We didn't have much, but it was so clean in our house that you could eat off the floor."

"Dad drove a truck for a living; he wasn't around a lot. He did mostly long hauls. Like your grandpa George, he was a strict disciplinarian. We did a ton of chores, and if we didn't do them, we got punished, usually with a belt, or they made us go cut a switch from a tree. There was no such thing as an allowance in our house. My father drank too much. Sometimes he would get violent, and my three brothers would have to get him under control. Despite that, he never drank on the job or missed a day of work in his life."

"If you messed up, punishment was swift and severe," my father added.

The more I heard, the more astounding it was to me that they had escaped. Somehow their morale and determination, unlike so many other children, had been strengthened, their resolve hardened by the destruction around them.

"My biological mother left us when i was five years old," my father said. "Sometimes when I got off school, I would see her dancing in bars as I walked home. I didn't blame her much for leaving, George was a hard man to live with, and she was a happy-go-lucky person. She couldn't take it anymore. My Dad had a hard time showing emotion and being around other people. He was practically a shut-in. My brother Virgil was an all-state running back for Charleston High and my father never went to a single game—not one. For the homecoming game, I had to go out on the field with him when he was announced at halftime because my father wouldn't come. I sang in several talent shows, and he never made it to any of those either."

That made me reflect on how things were in our family. My brother and I played sports, sang in the choir, performed in plays and talent shows in school, and my parents never missed a single show or game. They were our biggest fans, always. I thought about how hard it would be to get by, much less thrive, without their full support. My parents were always there for us. They had no problem telling us how much they loved us or how proud they were of us. Not a single day went by that they didn't hug us and say, "We love you."

"Your grandpa was one hell of a musician. He could play the piano and harmonica like nobody's business."

I reached into my pocket and felt the cool metal sides of the old harmonica my grandfather had given me before we left his house. After playing a few licks, he held it up, and said, "I will give this 'tin sandwich' to you, if you promise you will learn how to play it. I don't really use it anymore."

Of course, I vowed solemnly that I would learn to master it. How hard could it be? Just hold it up to your mouth and blow. I saw myself as another Huey Lewis, one of my favorite musicians, playing "The Heart of Rock and Roll." With Huey's inspiration, I believed I would learn to really wail on that thing—but of course I never did.

"The thing about your grandpa is that he was always scary smart. He read at least a book a week. He loved cowboy novels; Zane Gray westerns were his favorite. And he loved mystery stories. But really, he would read just about anything he could get his hands on."

Turning her head to look back at me, my mother said, "They didn't have a lot of schooling, but none of your relatives were afraid of hard work. Hard work is all they ever knew. They didn't have much, but they were resolved to never take a handout. They were far too proud for that. They didn't believe in taking charity from anybody, not like other country people who seemed to have given up entirely. At the same time, I don't think any of them believed that they would ever get out of the hollers, no matter how hard they worked. I'm not sure any of them would leave even if given the chance. I think some of them just accepted it as their fate."

I added somberly, but not believing my own words, "I guess you can't outwork fate." I was trying, but failing miserably, to find a silver lining in the family saga.

Dad went on, "As bad as we had it growing up, it was worse for your grandparents. Their parents worked 12-hour shifts deep in the ground mining coal. It was dark when they made their way into the mouth of the mine, leading into the bowels of the earth, and it was dark when they came out. They worked seven days per week, and they never saw daylight—literally. In those days, the workers made very little money, while the mine owners and operators got filthy rich off their sweat and suffering. They paid the miners in scrip, a form of currency that could only be spent at the company store. The shacks those families lived in were owned by the mining company as well and rented back to the mining families. So, they were forced to give most of their money right back to the company. It was a shit deal, and a brutal life. Most of the miners died prematurely of black lung disease, or as the result of industrial accidents. You were lucky if you lived to see 50. Many of our people didn't.

"Your grandpa George didn't finish school. He was pulled out and put in the mine when he was in fifth grade. He persevered there through his teen years. He was still young when his father died; his mother remarried a man with the last name Matson. Then his mom died, and he was raised by his stepfather."

In other words, George and those around him lived like indentured slaves. *What a miserable existence,* I thought to myself. I couldn't even fathom what that would be like.

"What was the worst thing about growing up in that little shack by the tracks?" I asked him.

"The rats would squeeze their way between the small spaces where the walls connected to the floor. It was especially bad in the winter when it was bitter cold outside. We didn't get fresh milk; it was too expensive. Instead, we had evaporated instant Carnation Milk that came in large tin cans. We cut the tops and bottoms off the cans and folded them at a 90-degree angle down the middle. Then, we would nail the bent circles of the cans to the bottom of the floor and side of the wall to keep the rats from coming in."

Part of the gloom and destruction of the town could be attributed to the presence of Union Carbide. My father talked about the smell of the chemical plant coming in through the windows at certain times of year. Chemical spills killed thousands of fish. The company's malfeasance was responsible for polluting a very beautiful part of the country.

Union Carbide built its first chemical plant in the country town of Clendenin, a comparatively short 30-minute drive from Charleston, along a winding road that followed the Elk River. The plant was created to produce ethylene, and it launched the petrochemical industry. Union Carbide would later cause what many claimed is the worst industrial accident in America's history. Between 1927 and 1932 they obtained the rights to drill a tunnel with the aim of creating hydroelectric power as part of an effort to end the Great Depression. The accident became known as the Hawke's Nest Tunnel Disaster. The tunnel itself was designed to be three miles long and drop a total of 170 feet, producing tremendous pressure to drive the turbines that would generate electricity.

An estimated 3,000 men worked on the project. During the drilling, crews encountered veins of almost pure silica, which can be mined for the electro processing of steel. Mining silica was extremely profitable. But in a direct violation of the regulations of the day, the men mining the substance were not instructed to wear masks. They freely breathed the deadly silica dust into their lungs. Eventually, Union Carbide would admit to causing 109 deaths as a result of ignoring regulations to protect the workers. Later, Congress declared the number of avoidable deaths was 476, and a final estimate put the number closer to 750.

Silica is commonly found in sand and quartz. When inhaled, it causes occupational silicosis, resulting over time in coughing, shortness of breath, constant fatigue, fever, swelling of the legs, and an inability to get enough oxygen into your lungs. There is no known cure. It is a form of slow torture as the lack of oxygen to your organs leads to a slow, painful, and terrifying death.

This travesty would pale in comparison to the worst industrial accident the world has ever seen, also caused by Union Carbide. In more modern times, the company is credited for manufacturing plastics, industrial gases, metals, medical products, paint, coating and wire cable, domestic goods, and far too many other products to list. Undoubtably, many of these products improve the quality of life for millions, if not billions, of people. At the end of World War II, they aided in the creation of the first atomic bomb, which is largely credited with helping bring the war to an end. Indeed, if the Germans had created it first, the world today would be a very different place.

Paradoxically, in 1984 this same company is said to have caused an incident that would become known as the Bhopal Disaster. Union Carbide owned 50.9% of a chemical plant in India. An estimated 45 tons of the gas methyl isocyanate escaped and poisoned thousands of innocent human beings. It is believed the incident was the result of faulty safety protocols and malfunctioning equipment. In efforts to cover up their culpability, the CEO at the time hired a PR company and began an aggressive cover-up, based largely on victim blaming.

Eventually they would settle the lawsuit resulting from the event, for $470 million, but not until 3,928 deaths had been certified. The averaged pay out for the deaths was under $120,000 per person. In the aftermath of the tragedy, the Indian government sought the extradition of the company's CEO to face murder charges. He remained safely out of reach in the United States and died 30 years after the accident at the age of 93. His name was Warren Anderson.

The environmental devastation caused by the plant has still not been cleaned up; mercury levels there remain six times higher than US standards—a grim reminder of the innocent people who lost their lives. This was the company my grandfather went to work for in 1945. But there were few options available to him to support his family.

"Dad, tell me about where Grandpa worked," I asked him.

"He worked out on Blain Island." That was an island just over a mile long, and about 900 feet in width. The island, I would discover, sits in the middle of the Kanawha River. Union Carbide bought it in 1921 and began producing petrochemicals. It would eventually be given the name Carbide Island by the locals.

"He mainly worked night shifts. After mom left, my brothers, sisters, and I were left alone at night. With no real supervision, we would run the streets. Some of the neighbors complained, and social workers put my sisters in a state-run facility for girls.

We went there every weekend to visit. I missed them terribly, especially Evelyn. We were very close. But at least they were safe when dad worked."

"How many people worked there?"

"It had to be thousands. At one point it was the number one employer in West Virginia. The pollution was so bad that the ash from the factory would eat through the paint on dad's car. There was an explosion at some point that blew my father over a six-foot fence. In the summers, my brothers and I fished the river by the plant. We caught some monster catfish and carp. But we never ate any of the fish we caught. We could see the red and green chemicals flowing out of the culverts on the island. I have no idea what that sludge was, but we knew it wasn't safe to put in your body."

The coal mines and chemical factories that seemed to run the state were polluting the land. Nonetheless, so many people chose to submit, stay, and suffer. It wasn't just the environment that was being contaminated—it was the mindsets of those who called this place home. It's as if they were caught by an invisible gravitational force and being sucked into a black hole of grief and destruction, unable to break free.

"At least our family was out of the coal mines," my father recalled, "which was way worse than factory work. Union Carbide was good to us kids. Every summer they would let us go to Cliff Side for two weeks—a kid's camp where they taught us how to swim. We were allowed to fish and hunt up in the hills. That's where I learned to shoot a .22, going after squirrels. At Christmas, the company would give us a ham. They gave my father an opportunity to go into middle management and be a foreman and have his own crew. They offered him a shot at getting more education and training." My father drew in a deep breath then let out a sigh. "It was our opportunity to get ahead. But he turned it down.

I was nonplussed. "He never took the promotion? You really needed the money. Who doesn't take a raise to help their family?"

"To his way of thinking, he would be betraying his heritage. 'I will never be the boss man and take advantage of other people,' he told me. 'They can keep their job; I am fine where I am.'

"Dad, that doesn't make sense. He could have helped the people he managed get better opportunities. He could have given his family a better life."

"When he told us about it as kids, he was proud he turned it down. But I was thinking pretty much the same things you are right now. We weren't allowed to talk about money, and we were never permitted to question him or his thinking. You would end up with a belt 'whippun.'"

I sat in the back seat struggling in silence to understand. George could have gotten off the line and worked in a much less toxic environment that didn't threaten his health and lungs. He was an intelligent man. He might have even risen to an executive level. He was hard-working and dedicated to the company. He had given 25 years of his life to Union Carbide, and when he had the chance to make his life better, he chose the lesser path. It boggled my mind. I felt inept and stupid the same way I did every time I picked up a Rubik's Cube and half-heartedly tried to solve it. While Brian and Matt both mastered putting all the colors on the right sides of the square, all I ever ended up with were multicolored sides and frustration. The same disturbance I felt now. Unlike the cube, I resolved to make solving this conundrum part of my life's mission.

We talked in the car for nearly three hours. As we got closer to home, Mom turned the radio on. After a few minutes of channel surfing, she found our local country station, and the sounds of John Denver's "Country Roads Take Me Home" filled the car.

As the song continued, we all started to sing together. It wasn't the wild and crazy "Margaritaville" singing. It was a somber, thoughtful, and grateful type of singing. The kind of singing that fills you up and somehow empties you out at the same time. The kind of singing that is real and true, like a hymn.

I was indebted beyond all measure to my ancestors who had braved and survived the hills, coal mines, and chemical plants. Like John Denver, I was proud to be a country boy. But although I loved the song and I loved John Denver for writing it, I knew I didn't belong there. My mom and dad had probably saved my life by getting all of us out. Whatever my fate held, I would not find it in the hollers. There was a heavy weight of anguish, resentment, and resignation that, like a spectral ghost, hung over that place. It was a mood of decay and profound helplessness. For far too many, it was victimhood mixed in with the unholy cocktail of sloth and entitlement.

I cried at the end of the song, but these were tears of gratitude and love. I had escaped with my parents. We were a long way from wealthy, but we had something much more valuable. We had optimism, and a burning drive to build a better life. We loved our country and our lives. We cherished and honored those who had died to give us our freedoms. We had the firm foundation needed to build the American Dream. It was the greatest gift my parents could give me. Somehow, my parents were able to "see further" and imagine more. They were able to extricate themselves from the self-destructive mindsets of our forefathers.

And it hit me like a bolt of lightning. My father and grandfather had not lived in the same world. It was like they existed in two separate dimensions. It wasn't just that they saw the world differently. Their mindsets and the screens they lived by had created two separate realities. In effect, they didn't even speak the same language.

Modern physics makes the claims that there may be many dimensions we cannot see or have access to directly. When it comes to people, the human mind can create a multitude of dimensions, some of which give you extraordinary power to act and live an adventurous life, and others that disempower and imprison us.

And if that were true, it might be possible to study the distinctions of varying worlds and choose powerfully the ones that best serve you.

Several years later, my grandfather and Retha made their one and only trip to Greenhills to visit us. Dad was eager to share the life and the home that he and my mom had made for our family. It was a modest home of 2,000 square feet, with a small pool and a grill in the back yard, but we kept it immaculate. Somehow, it made all of us feel loved.

The day they arrived, it was cool for August, and the humidity was unusually low, so we had the air conditioning off. The screen door let the light breeze in through the front of our home. There was no smell of death here. We heard their car pull up, and we all hurried to the front door to greet them.

George led the way. As he pulled the screen door open, he said to my father, "How many people did you have to rip off to get this place?"

He didn't see the love, hard work, and dedication that my mother and father had poured into this world they had built for me and my brothers. The idea that my parents helped people was beyond his comprehension.

He verbally attacked my father in front of his whole family, accusing him of being a crook. My father kept his cool. He said, "George, Mary Lou and I are working hard for our family. It's difficult work, and there is risk involved when owning your own business. The way we see it, when you own your own business, the more people you help, the more you make. Not everyone who earns money is greedy or a criminal."

George fired back, "Believe whatever you want if it helps you sleep at night. But what I see is that my son thinks he is some kind of big shot. Is that what you are, a big shot? Do you think you are better than me? Are you better than me now that you have this big house?"

At that point, the argument exploded into a full-blown shouting match. A few minutes later, George was back in their car, headed home to the West Virginia hills. But in a way, he brought that world of resentment and resignation with him everywhere he went. As my grandparents drove away, we stood in our living room as if cemented to the floor, unable to move. Matt, who always brought a sense of comic relief to our family in time of stress, said, with a deadpan face, "Well, that was messed up." We all laughed hysterically. My grandfather had seen the physical structure that we lived in, but he was blind to the power of our family, and the love that bound us together.

My father and grandfather could not have been more different. George had done the best he could for his family, but self-expression and intimacy eluded him. From his perspective, the world was full of danger and hate. It was a place to insulate yourself from, a place to be dreaded and feared. Those beliefs caused decades of anxiety and depression for him and his family.

My dad forged another path. He saw the world full of opportunity and adventure. He liked people, and he liked life. He trusted people, and he never gave up the idea that he could make the world a better place. Dad passed that down to me. He was my hero. Not just for what he had achieved in the world, but because he always believed in me, and he made sure I knew it. He made sure we all knew it. I could count on both of my parents to support me and build me up. Even when I was young, they believed I had the power to accomplish amazing things.

George had kept his family alive, but it was my dad who gave us a life worth living. For my dad, the American Dream was real, and his to claim. My grandfather did not believe in the American Dream. In fact, there is a good chance he despised the very idea of it. George saw my father's success as an indictment of his own failures. My grandfather came to visit us that one and only time. I know my father wanted him to be proud of what my parents had accomplished. But it was a bridge too far for George to cross. He could not acknowledge or be part of my father's achievements. He wanted nothing to do with them. And that was the last time any of us would see George.

George died alone in his small house in Charleston, of emphysema, contracted from all those years in coal mines, and living in houses with fireplaces lined with asbestos, not to mention the toxic chemicals out on Carbide Island.

And from a young age that got me thinking: what are the screens through which we see the world? Why did my family have such a different outcome than my grandfather did and what does this all mean for an understanding of the American Dream?

Introduction: The Way Ahead

"My invitation to you is to begin living every moment as though you are miraculous and deserve to live an extraordinary life. . . . The gift you will be giving yourself is a lifelong journey of discovery, one that is infinite and infinitely rewarding. Begin the journey. Today. This moment. Now."

—Robert White

Many people are losing hope in the American Dream. The simple idea that anyone can get ahead in America if they work hard and apply themselves is under attack. Although it is hard to put a specific date to it, the ideas of free markets, capitalism, and self-accountability leading to a fulfilling life have been dealt a black eye by some people who believe in entitlement, victimhood, and collectivism. The years since 2020 have been financially and emotionally difficult for many people. After the COVID-19 pandemic, a recession followed, and that led to an uptick in violence and political division. Today it's all too easy to feel that America is going in the wrong direction and any real opportunities to grow personally and financially are drying up. But let me cut to the chase: the American Dream is still very much alive for anyone willing to take responsibility and fight for what they love. You are probably one of those people or you wouldn't have picked up this book.

While I will make the case that the American Dream is yours to create and that it can lead to an extraordinary life, it would be foolhardy not to address the challenges we have had and are now dealing with. To most people it seems we have faced one form of chaos after another since early 2020. The first COVID case in the US was diagnosed on January 20, 2020, in Snohomish County, Washington. Globally, we are still reeling from the destruction of the virus.

The loss of life and the toll it took on our physical and mental health as a society are difficult to overestimate, not to mention the financial impact it had on our global economy. During the shutdown in the United States, businesses went under, and school closings hurt our children's ability to learn, grow, and experience much-needed socialization. In an attempt to deal with this medical and financial emergency, the United States government racked up massive debt in the form of deficit spending. To make matters worse, the spending continued long after the pandemic seemed to subside, resulting in the emergence of massive inflation. Many families struggled to pay their bills. Savings rates dropped, and many people took out loans or withdrawals from their 401(k) to make ends meet, thereby depleting much-needed funds for retirement.

Many of our larger cities have experienced violent upticks in crime and homelessness. Drug use has fueled the homeless crisis as tent cities continue to spring up. As many Americans struggle financially, war has erupted. Russia invaded Ukraine. Then, Hamas launched a ruthless and inhuman terrorist attack against Israel causing yet another war in the Middle East. Simultaneously, many Americans are concerned about the massive numbers of illegal immigrants crossing into this country, often fueling sex trafficking, drug smuggling, and increases in potential terroristic threats. I am not asking you to turn a blind eye to these realities or to downplay their tragic effect but to consider that the American Dream is still available despite them. Indeed, it may even prove critical to addressing many of these tragedies head on.

If you feel like throwing in the towel and giving up on your dream, you wouldn't be alone. In a 2023 *Wall Street Journal* poll,[1] only 36% of respondents believe the American Dream still holds true. That is down from 68% just a year prior. Half of all people in the survey felt that America is worse off than it was 50 years ago. Fifty percent strongly or somewhat agree that "the economic and political system of the country are stacked against people like me." The survey didn't identify what "people like me" means, although 39% strongly or somewhat disagreed with the same statement. A 2023 survey conducted by NBC[2] reported that only 19% of people felt their children's generation would be better off than the current one. Even more shockingly, a survey by NORC at the University of Chicago[3] claimed 18% said the American Dream *never* held true. Emphasis added.

Considering the recent past, few people feel optimistic about the political and economic realities of today. The fear of war alone can be a daunting emotion. When coupled with worries of inflation and recession, life can seem outright bleak.

Many people view the American Dream as a purely financial endeavor. Sadly, some believe the American Dream as a form of unbridled materialism based on greed. Although it does have a financial and investing component, the American Dream is not about money or wealth. Money is fuel for the dream, but not the dream itself. The American Dream is not something you own, capture, or get—it's a way of being. It's a form of expression. More specifically, it's a mindset or screen through which you see and comprehend the world. And when viewing the world through the American Dream screen you will experience a new sense of empowerment. It will transform your idea of what is possible in your life regardless of the external circumstances of the world.

As I help you create a new perspective for your American Dream, we will inevitably need to examine how you invest your money to fulfill and expand on what matters most to you. After managing billions of dollars for investors over the last 40 years, I know the inside baseball. I've seen the dark and dirty side of this industry, and I was also unsuspectedly part of it. I will show you how that game is played. I promise not to sugarcoat anything. I will tell it to you like it is because you deserve to know the unvarnished truth. It might make you uncomfortable, and it might not be what you want to hear, but you need to hear it anyway, because the truth has power.

When it comes to putting their hard-earned money to work, most people want simple hints and tips to make money and accumulate wealth. The problem is that hints and tips don't work. I could easily write a concise, do-it-yourself investing pamphlet about constructing a portfolio using asset allocation. I could suggest ways to build your portfolio, and how and when to consider rebalancing it.

I can tell you what asset categories might help maximize expected returns and what investments you should avoid. If you were pressed for time, I could probably give you a brief rundown in five minutes. Heck, I could even draw a pie chart for you and label the pieces you might buy. However, there are many of these books already out there. The problem is they offer simple, therefore flawed, answers to complex questions. This is not a quick and easy, do-it-yourself book.

Most people think they only have two options when it comes to investing: do it yourself or hire someone to do it for you. There is a third option that most people don't realize they have: working hand-in-hand with a coach to train and develop you to make the right decisions. In this book, I will lay out the academically proven investing principles my clients have benefited from over the past three decades. However, that information alone will not help you.

This has nothing to do with you. That information won't help anyone because human beings just aren't engineered to properly invest and maintain anything like discipline. We may think we're prudently investing, but what we're really doing is speculating and gambling with our money. That's essentially what stock picking, market timing, and track record investing are: other forms of gambling. Even if we're soundly investing and doing well for a period, what happens when the S&P 500 suddenly drops 50%, as it did back in 2008 and 2009? What happens if you turn on 24-hour cable news and the host is frantically making predictions that create uncertainty and scare the heck out of you? If you're like most people, chances are high that you'll panic, sell low, and lock in your losses. Why? It's human nature.

That is not all you have to deal with as a modern investor. What happens when some new exotic "investment" vehicle like cryptocurrency seems to turn people into multimillionaires overnight? Or what happens when all of your friends are jacked up about how much money they are making on individual tech stocks or on exciting AI companies? If you are like most investors I know, it stimulates fear. But not the fear of losing your money; this is the fear of losing out on high returns everyone else seems to be easily raking in.

In the world of investing, it is not what you know, but what you do that counts. This isn't only true about investing. It's true about anything in life that is difficult and requires discipline over a long period of time to be successful. Look at the diet and exercise industry. By this point, almost everyone understands what it takes to lose weight and live a healthier life. It requires eating less and moving more. Pretty simple. But do most people do it? You might be in the category that does, and I sincerely hope you are. Unfortunately, the fact that more than 41% of Americans are obese says most people don't follow these simple rules.[4] We know junk food is bad, but people still eat it, and I am no exception. Most people who smoke cigarettes know that it's terrible for them, but it doesn't stop them. If being healthy and getting fit came down to knowing the information, we'd all be back down to our high school weight and 15% body fat.

We have the same problem when it comes to money and investing. Everyone knows what it takes to save money. It requires spending less and saving more. Do most people do it? Of course not! According to a wellness survey conducted by First National Bank of Omaha, nearly 46% of Americans have less than $15,000 in savings, and 59% fear they won't be able to retire by age 65.[5] Fifty-seven percent of Americans can't afford a $1,000 emergency expense, and 68% are worried they couldn't cover a single month's worth of living expenses if they lost their primary source of income.[6] This is supposed to be the most prosperous country in the world, but far too many people live paycheck to paycheck. The information about how to avoid these situations is out there, which proves that information alone isn't enough. That's why hints and tips don't work.

Becoming successful in harnessing the transformative power of the American Dream goes beyond discipline. It requires change. The groundbreaking book *Change or Die* by Alan Deutschman proves that most people refuse to change even when they face serious health issues that will literally kill them. In these cases, people are "dying" to stay the way they are instead of taking on a new way of being and a new set of actions. The same can be said of companies that need to evolve to survive. More than 70% of all change initiatives fail for companies.[7] Given that the pace of social, economic, and technological change is exponentially speeding up, the ability to change when old paradigms no longer work is critical. I often tell our team members, "'It has always been this way' is where the future goes to die." Your success in achieving and elevating your American Dream will be directly related to the actions you take and the actions you eliminate. Change will be critical, full stop.

This book will also not be about setting goals. There are already far too many books on how to set and achieve goals. There is nothing wrong with goal setting, per se, but goals do not have the ability to fulfill your American Dream or any other dream for that matter. Think of a time when you set a financial or investing goal. Perhaps you worked hard and bought that new car or your dream house. It felt good for a while, but after that exciting feeling wore off, you needed another goal to pump you up again or keep you going.

I will help you think beyond achievement of short-term financial goals of making more money so you can purchase more material goods. I have nothing against material wealth, but unfortunately that alone will not leave you feeling alive and inspired. It won't give you a life you love. Once you make more money or go out and buy that fancy new item, you often realize its effect is only temporary and fleeting at best. Those financial goals have the funny effect of leaving you feeling empty after you achieve them, so you need to go out and set a new goal. You can make more money and buy more things, but no matter how much you accumulate, it doesn't get you where you want to go. Sometimes, it can have the opposite effect.

Early in my career, I discovered that some of the wealthiest clients I worked with were the unhappiest. It didn't matter how much money some of them had, they still panicked every time a cable news host spread fear about the market or chased whatever hot fad seemed to make other people boat loads of cash. Others have let their wealth create tension that has damaged their relationships with friends and family because they had an (often unspoken) "no-talk rule" regarding money. I've watched money tear families apart. Why? Because money can create jealousy, resentment, and division. Suddenly, even wealthy people behave as victims and look to lay blame elsewhere for their troubles. What good does all that money do if it creates misery, doubt, and fear? You don't need to look far for examples of people with massive wealth that experienced suffering and often self-induced pain despite their riches. Consider Elvis Presley, Howard Hughes, or Marilyn Monroe, just to name a few.

There is nothing wrong with achieving and accumulating wealth, even massive wealth. In fact, if you create enough value for other people, you are likely to achieve immense financial success. But that alone will not ensure an extraordinary life: a life that is unique, remarkable, and inspiring.

That's why if you go into this book simply hoping to improve your portfolio or earn enough money to secure a lovely vacation home in Boca—and I'm not criticizing that—you will not receive the full benefit. The real value of this book goes much deeper. This book is not about getting anywhere or acquiring material possessions; it's about transforming who you are and how you see the world. It's about attaining an extraordinary life through your American Dream, regardless of your life circumstances. I'll take you step-by-step through the process.

Part I is about identifying what's most important to you so you can establish your true purpose for money and your life. In the absence of purpose, no amount of money will ever be enough, but when you have purpose, you will learn that you won't need money to experience fulfillment. When it comes to this type of transformational change, the information alone will only take you so far. You need purpose.

Part II will point out the challenges and roadblocks that stand between you and your American Dream so you can better understand the aspects of the financial industry and your own human tendencies that are working against you. With that understanding you can reap the benefits that money and the science of investing may provide.

In Part III, I will introduce you to a whole new world of investing that you didn't know existed. If you commit to following these academically tested, Nobel Prize–winning investing principles, you can finally stop what academics view as gambling and speculating with your future. I will give you access to the tools you need to make powerful financial choices for your family. You will not only discover what intentional investing looks like; you will also walk away from this book feeling empowered to create an investment strategy aligned with your personal purpose.

More than anything else, you will discover what it truly means to live an extraordinary life empowered by freedom, fulfillment, and love. Freedom from the internal and external constraints that prevent you from expressing yourself. Fulfillment in the form of joy and excitement about your life, so you feel engaged and inspired. Love as a connection and closeness with your family and others. If you can do that, you will have an extraordinary life that no amount of money or status could replace. That is the American Dream.

I've boiled this process down to seven actionable life strategies that everyone can understand and follow.

1. Ask powerful questions.
2. Develop the screen of the American Dream.
3. Transform your relationship with money.
4. Establish your true purpose for money.
5. Create a powerful future view.
6. Take massive action.
7. Embrace academic investing principles.

These strategies fit together and reinforce each other. But for this process to be effective, you (and only you) are responsible for putting in the work. No one can do it for you. It is a lifelong journey. Although investing will always involve risk, following the strategies in this book will put you on the path toward investing peace of mind. It can help ease the pain, suffering, and confusion so many people have about money. For far too many people, money shows up as hard to make, hard to keep, and hard to invest. You will discover it doesn't have to be that way. You can alter your relationship with money so that it is no longer a disempowering force, but an inspirational one that feeds your dreams.

When I first contemplated writing this book, my agent asked me, "Who exactly is this for? What is the target market?" At first, I struggled to answer this. I couldn't pin down a single demographic or specific type of person. What I do know is that when I train and develop people during the American Dream Experience workshop, we attract people from all walks of life. We have people enrolled in the program who are just starting their professional careers and are new to the world of investing, and we also have successful people who have already amassed millions. There are also experienced entrepreneurs and innovators, those getting ready to retire, and even those who have built successful businesses and accumulated wealth and are ready to start a new chapter in life.

These people report breakthroughs in their lives, if they are willing to work for it. So, who is this book for? Anyone and everyone may benefit from the American Dream and the principles laid out in this book. But before we talk about money and investing, we must first address human behavior, how you view the world, and your place in it. I'm going to warn you ahead of time: this will require you to change your thinking, so if you are willing to grapple with old beliefs and ready to fight through cognitive dissonance when it comes to change, this book is for you. Let's begin.

Notes

1. Aaron Zitner, "Voters See American Dream Slipping Out of Reach, WSJ/NORC Poll Shows," *Wall Street Journal,* November 24, 2023. https://www.wsj.com/us-news/american-dream-out-of-reach-poll-3b774892.
2. Mark Murray, "Poll: Biden's Standing Hits New Lows Amid Israel-Hamas War," *NBC News,* November 19, 2023. https://www.nbcnews.com/politics/2024-election/poll-bidens-standing-hits-new-lows-israel-hamas-war-rcna125251.
3. Zitner, "Voters See American Dream Slipping Out of Reach."

4. "The State of Obesity: Better Policies for a Healthier America," Trust for America's Health, September 2022. https://www .tfah.org/wp-content/uploads/2022/09/2022ObesityReport_ FINAL3923.pdf.

5. Sarah O'Brien, "Saving for Retirement Is the Top Financial Priority for Just 17% of Adults, Survey Shows," CNBC, March 31, 2022. https://www.cnbc.com/2022/03/31/saving-for-retirement-is-top-financial-priority-for-just-17percent-of-adults.html#:~:text=Your%20 Money-,Saving%20for%20retirement%20is%20the%20top%20 financial%20priority,17%25%20of%20adults%2C%20survey%20 shows&text=Nearly%2060%25%20of%20respondents%20 in,saved%20for%20their%20golden%20years.

6. Lane Gillespie and Tori Rubioff, "Bankrate's 2023 Annual Emergency Savings Report," Bankrate, June 22, 2023. https:// www.bankrate.com/banking/savings/emergency-savings-report/#n3i.

7. Nitin Nohria and Michael Beer, "Cracking the Code of Change," *Harvard Business Review*, May–June 2000, https://hbr.org/ 2000/05/cracking-the-code-of-change#:~:text=The%20brutal %20fact%20is%20that,an%20alphabet%20soup%20of%20 initiatives.

A Whole New Way to See the World

"Create your future from your future, not your past."

—Werner Erhardt

"What are the screens through which you currently see the world?"

When I was eight, I asked my dad, "How did you get out of West Virginia?"

He was getting ready for work, tying his tie in the mirror—it was always a double Windsor, that was important. "It's a matter of asking the right questions," he told me. I didn't get it. That sounded like something Yoda would say in *Star Wars*. "What do you mean?" I asked.

"If I only asked myself, 'Why am I doomed to live my life in the hollers of West Virginia?' my life would turn out one way. But because I asked the question, 'How can I get out of the hollers to pursue the American Dream and create prosperity for my family?' my life has taken a completely different trajectory. You will always search for the answers to the questions you ask, so ask the right questions, and the answers will empower you to be curious about the world. That allows me to see things other people miss."

Most people don't want to change their perspective, so they don't bother to ask challenging questions. They aren't curious about the world because they believe the world has always been and always will be a certain way. If they ever do ask questions, they want life hacks. They want to know how to get what they want in life as fast and with as little effort as possible. That's about as deep as it goes for most people. But in my experience, those quick, easy answers don't make a profound difference in your life. The difficult questions—the ones you consider and dwell on for an extended period—have the most impact.

So, at the beginning of each chapter of this book, I will pose a question. I don't want you to answer it right away. I want you to sit with the question and think about it for an extended period. And not just when reading this book. Ponder it whenever you get the opportunity. Galileo, Einstein, and all the great thinkers throughout human history asked profound questions they pondered for decades, sometimes even their entire lives. Einstein famously said that he spent 95% of his time on the problem and only 5% of his time on the solution. I want you to focus on the problem at the root of each question and not rush to find an answer.

It sounds easy at first, but you will encounter resistance because your brain will want to call on the information it already knows to find a quick answer. Instead, resist the easy answer, and dwell on the question because, on closer examination, you might realize you don't have the complete answer. Don't believe me? When you get further along in the book, refer back to some of these questions from previous chapters to see if your answers have evolved or changed entirely. That will help you better understand and expand on your thinking. When you allow yourself to remain in and live with profound questions, that's when true breakthrough thinking occurs. That's how you learn what you don't already know.

Here's what I mean by breakthrough thinking. This circle represents all of the potential knowledge in the universe (see Figure 1.1). Yes, there's a lot in that circle.

This piece represents everything that *you know* you know (see Figure 1.2).

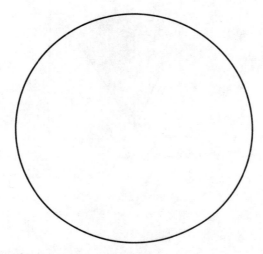

Figure 1.1   All Theoretical Knowledge

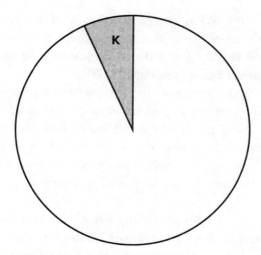

Figure 1.2   Theoretical Knowledge You Know

You know what a stock is. You know what the sun is. You know your children's birthdays. You know who the president is. Whatever it is you know, it fits into this piece of the pie. Keep in mind that this isn't drawn to scale. Think of all of the possible knowledge available

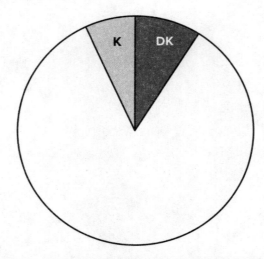

Figure 1.3   Theoretical Knowledge You Know You Don't Know

within the universe. What any of us know is probably a speck, but I will give you credit for this entire piece for our purposes.

This second piece of the pie represents all of the things that you know you *don't* know (see Figure 1.3).

You probably know there are nuclear reactors, but most of us don't know how they work. You know of airplanes and aeronautics, but unless you're a pilot, you don't know how to fly an airplane, never mind how one works. There are plenty of things out there that we know we don't know. This piece can only be informed by the previous piece (what we know we know), which is why it's so limited.

When learning about investing, most people want to take everything they know they don't know about investing and move it into the piece of the pie that represents what they know they know. If they can do that, they believe they will become more successful investors. This is where the hints and tips come in. Most people understand the basics. They might even be happy with how they're investing but want some of those hints and tips to increase their returns by 1% or 2%. However, that won't get you where you

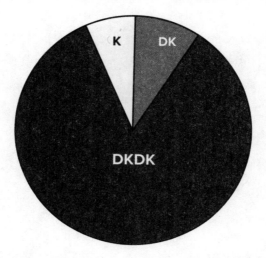

Figure 1.4   Theoretical Knowledge You Don't Know You Don't Know

need to go. It keeps you trapped. To experience true breakthrough thinking, you must learn what you never considered possible.

This remaining portion of the pie represents all of the things you don't know that you don't know (see Figure 1.4).

This piece includes subjects that you don't even know exist. I know it sounds daunting. No book in the world could contain all of the things you don't know you don't know, so I'm going to focus on two primary areas.

The first is academically, empirically tested research about investing. The second is how your brain works. So, in addition to investing, you also need to learn things you don't know you don't know about yourself. It's in these two areas, the science of investing and yourself, that the majority of this book will focus on.

The single best way to achieve prosperity and the American Dream is by delving into and allowing yourself to linger in the slice of the pie that is what you don't know you don't know. You want to remain in this area so you can explore and discover new possibilities. You do that by asking questions and being curious about the world, how it functions, and your place in it. This is a lifetime

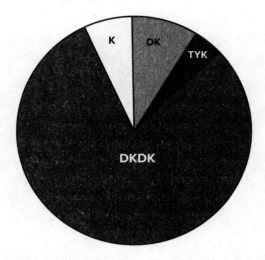

Figure 1.5   Theoretical Knowledge You Think You Know but Is Not So

pursuit, and if you commit to it, it will make all the difference. But that's not everything. There is still one area that we have to account for (see Figure 1.5).

This nasty little slice over here represents the things you think you know but are not so.

This is where people get hung up, for it's in this slice where resistance and cognitive dissonance occur.

Fighting Through Cognitive Dissonance

When you believe something and have been operating under the assumption that it is true (whether it is or not), your brain will make every attempt to prove that it is.

It doesn't matter who you are; when you learn that what you thought was true is not true, something called *cognitive dissonance* happens in your brain. That is the anxious and uncomfortable feeling that comes from having inconsistent thoughts, beliefs, or attitudes, especially relating to behavioral decisions and attitude changes. This can also affect you physically, causing you to sweat, feel flushed, and even get sick. It can be an incredibly uncomfortable feeling. Whether you came to the original belief from family, friends, television, or wherever, it's natural human instinct to resist that new way of thinking.

We are so emotionally invested in our beliefs that we cannot recognize them as anything but the truth. We also have a nasty habit of falling in love with our beliefs. That explains why being wrong can so easily wound our sense of self and is why we resist change and ignore evidence, even when it's compelling. It's why we try to alter those new ideas to make them fit with our existing ideas, thus preventing us from getting their full impact. Or worse, it's why we never change our thinking, keep doing what we've always done, and continue getting the same results. Think about a time when you changed a belief or attitude; chances are it wasn't the most comfortable experience. However, a closer look will reveal that it probably also led to a profound breakthrough. My intention is that this will happen again when reading this book. You will have to test your beliefs because it's difficult to accept that when you thought you were prudently investing, you were actually speculating and gambling with your money, or that your subconscious is calling most of the shots.

When experiencing cognitive dissonance, people tend to check out because they don't want to confront that they were wrong. They put the book down or change the channel. Anything to avoid that feeling that comes with cognitive dissonance. It doesn't help that there are short-term benefits you can experience by not challenging your beliefs. Psychologists call these *secondary gains*, and they reduce your ability to act. Consider the following list of "benefits" that can prevent us from embracing new ideas:

- **Being right.** It's just fun to be right. Who doesn't love it?
- **Avoiding being wrong.** Nobody ever wants to admit being wrong. It requires swallowing your pride and putting aside your ego. It is not a comfortable feeling, so we avoid it whenever possible.
- **Self-justification.** Just because things have worked out in your favor, that doesn't mean it's the best possible way to

do something. But those previous successes are an excellent reason your brain can convince you not to change. *That stock I picked went up 250%. Stock picking can't always be bad.* When you have that voice in your head, it's easy not to change.

- **Magical thinking.** You probably have a system, and that system has probably worked in the past, but did it work because it was a sound system or because the stars aligned and you got lucky? It's more comfortable to believe that we're good at this and know some things others don't when our success might have nothing to do with that.

- **Perceived control.** With so much going on in the world outside of our control, it's comfortable to believe we are in the driver's seat when it comes to our money and life. *I've always done it my way, and it's worked for me so far. Nobody needs to tell me what to do.*

- **Playing God.** Things go so well every once in a while that we assume we can predict the future. When you're on a roll, it feels like it will never end, and it becomes easy to get overconfident. *I've read up on it and done my homework. I got it. No problem.* It sure feels good to pretend, but that's playing God. Nobody knows the future.

- **Protecting my identity.** Sometimes, we are so used to doing things a certain way that it becomes a part of our identity, so to do things a different way would require a complete change of identity. *Who am I if I can no longer do things this way?*

- **Fear of missing out (FOMO).** We all know about and have experienced FOMO. *Everyone is talking about this stock. There is a huge opportunity here. What happens if I don't take advantage of this, and it ends up being big? I would kick myself if I didn't jump at this opportunity.* Maybe that's all true, but maybe not.

- **People pleasing.** Nobody wants to disappoint others, which sometimes results in us doing things we wouldn't usually do because we don't want to disappoint anyone. *I've been investing with my friend for 20 years; I feel I owe it to him. Besides, we're friends. It would destroy our relationship.* If that person is really your friend, they will remain your friend whether you invest with them or not. More on this later because it's much more common than you might realize.

- **Tax aversion.** I get it. I hate taxes, too; taxes are just one factor when considering whether a change is in your best interest. *I could prudently diversify my money, but I would have to pay extra in taxes.* Guess what? It could very well be worth it.
- **Minimizing problems.** *What I'm doing right now is not all that bad.*
- **Avoiding responsibility.** *I don't want to learn a new way of doing things.*
- **Regret avoidance:** *What if I do this, and it doesn't work out? Then it's my fault, and I'm the one who ends up looking bad.*
- **Looking good.** *What will people think of me if I start doing something new, even though I know it's prudent for me to do?*
- **Being smart.** Arguably, being smart is the only thing some people like more than being right. *If I figure out this thing that nobody else can, then I know that will leave a positive impression on people, so I must keep doing it my way.*
- **Fitting in.** When everyone is doing something, it can make us question our choices, even when presented with contradictory evidence. *Everyone is doing this thing this way, so if I suddenly started to do something else, I would feel like the odd one out.*
- **Addiction.** Technology has made destructive behaviors like stock picking and market timing a major problem. *I can't stop doing what I do. I'm hooked, so to do things differently would rob me of that rush I experience when buying and selling stocks.*
- **Avoiding being controlled.** *I won't let you tell me what to do. I want to be in control of my destiny. Even if I make mistakes, they are on my terms.*
- **Getting rich quick.** Like with hints and tips, we all want things to be easy. We want instant results without putting in the work. Some of the best-performing stocks in 2023 went up over 1,000,000%. It would be nice to know that in advance, right? But guess what? You have absolutely no way of knowing.
- **Fear of change.** Even positive change can be difficult if you're used to things a certain way.

Despite what you may feel at the moment, these aren't real benefits—they are emotional payoffs for doing nothing. It's justification for inaction. This type of reasoning will not improve your life, but these traps are so easy to fall into. This is nothing to feel shameful or guilty about because these are all natural, human instincts. They occur beneath the surface, on a subconscious level, and they are calling the shots. That's why it's so hard to change your thinking and actions, and this is what leaves so many people trapped, living an ordinary life. So, how will you respond if I present you with information or make an assertion based on scientific evidence that bounces up against or contradicts something that you always believed to be true?

I'm telling you this now because I want you to expect this feeling and recognize cognitive dissonance when it occurs. I want you to avoid a knee-jerk reaction, or any of the listed excuses that will prevent you from embracing this information. Instead, face it head on. Grapple with this feeling and try to better understand the new information. It won't be easy, but it's often the only way you can learn, change, and grow. However, cognitive dissonance is only the tip of the iceberg. There is much more at play that shapes your current view of investing, the world, and your place in it.

The Terministic Screens Through Which You See the World

To effectively combat cognitive dissonance so you can embrace new ideas, it helps to better understand your perception, specifically, the terministic screens through which you view the world.

In the *Three Laws of Performance*, authors Steve Zaffron and Dave Logan write, "A person's way of seeing a situation is filtered through what scholar Kenneth Burke calls a terministic screen. This screen comprises language, words, terms, phrases, and their relation to each other. This screen directs our attention to aspects of reality. We don't see the world, and we don't see the terministic screen. Instead, we see what the screen allows us to see."[1]

Think of a screen as a worldview formed over time by everything you've seen, heard, read, and experienced. We all have screens—many of them—influencing how we view the world and our place in it. They also influence our language and the stories we tell ourselves, so they can't help but influence our beliefs and actions. It's why what we do seems obvious to us and might confuse others. It's also why what others do and say might sound completely irrational to us. It's like we're all wearing a unique pair of contact lenses specially made for us. We've become so comfortable wearing these contacts that we forget they're even there, and most people have no idea they exist.

Screens aren't necessarily a bad thing. Doctors have a screen that enables them to see the world in a way I cannot. If someone comes to them with a certain ailment, they are much more qualified to diagnose and treat it because of their screen. No doctor was born with that screen, and it didn't appear overnight. It was created over time by accumulating knowledge and experience.

Screens are often created without our intention, but you can choose the screen that empowers you to take action and live an extraordinary life. However, you must choose your screen carefully because it will either influence you for a grand purpose that helps create fulfillment so you can live your dreams, or it will influence you for a negative purpose. The choice is yours. Screens can change your life. I know that's true because they have changed mine.

I spent most of my younger years as an atheist. Not agnostic—that was for wimps. I was a straight-up atheist. I went to church because I sang in the choir and liked to perform, but I didn't connect with religion, and I didn't want to believe it. *There is no God. People made up God to feel good about themselves and create false hope.* As far as I was concerned, it was all a bedtime story. Looking back now, I realize that I didn't want to have something else in control of my life because I felt that made me weak. While attending the Miami University in Oxford, Ohio, I was a finance and accounting major, but I also studied psychology and learned about all the problems those in the field had with God and religion. So, I rejected religion and chose to live in a world with no God. That was a terministic screen that shaped my world, and that screen would influence me.

In my late thirties, I went through a divorce. I moved out of the house and into a small apartment. I could only see my kids half the time, which took its toll on me. I felt like a massive failure and a terrible person. I was scared that my kids would be scarred for life. Without guidance, I spiraled down, and my world started falling apart. When at my lowest, one of my buddies, who was a Christian, asked me, "How does it feel trying to run your own life?"

"It doesn't feel good," I told him.

"Are you ready?" he asked.

"Ready for what?"

"Ready to hit your knees and pray the first honest prayer of your whole life? Are you ready to accept that there is a God and you aren't him?"

I looked around that tiny apartment and realized this was where a life with no God had led me. At that moment, all of the psychology classes I took, all of the beliefs I had, and all of the screens I had created vanished. Gone! I hit my knees, and together with my buddy, we said a prayer. I went from an empty and hopeless universe of my creation where there was no God to a completely different world of love and fulfillment with God at the center.

I'm not trying to recruit anybody. That's not what this book is about. I believe everyone is free to believe whatever they want to believe. However, I do know that everything I have in my life today—my marriage, business, kids, community—none of it would have been possible had I continued living in a world shaped by my old screen with no God. That's how powerful screens are.

I want to go back to the question posed at the beginning of the chapter: *What are the screens through which you currently see the world?* What thoughts and beliefs make up how you see the world and your place in it? Yes, I'm talking about money and investing, but also consider your life as a whole, because it's all connected. Some screens will be obvious, and others might take longer to recognize. Remember, when you see the world through a particular screen long enough, you fail to see the screen—just what it allows you to see. However, now that you know what to look for and better understand the framework, it will be easier to recognize. If you struggle at first, just keep at it, because it's not going to happen on its own. You do need to put in the work because you, and only you, are capable of creating your transformation.

Responsibility Versus Victimhood

"You have to come into my office right away."

Nobody wants to hear their doctor say that over the phone. "What's wrong?" I asked. "Do I have cancer?"

"No, you just need to come in."

I was 32 years old at the time, and in complete despair. When I arrived at the office, my doctor told me that I had osteonecrosis as a result of the prednisone I had been prescribed for asthma. That's not what I was expecting to hear. I didn't even know what it was. He explained how *osteo* means "bone" and *necrosis* means "death." The head of my femurs were destroyed.

My initial reaction was to ask for a pill to fix it. "Oh, no," he said. "It doesn't work like that." It turned out the pain was going to get much worse as my femurs degenerated, and I would soon need surgery, or be confined to a wheelchair for the rest of my life. The pain would be intense, but the surgery would be no picnic, either. They would have to cut off the head of my femurs, so they could drive a giant rod down into the bone, and then replace both hips.

I was devastated. I couldn't play soccer with the kids in the yard. I couldn't even pick them up. There was a period where I started to feel sorry for myself. I was in a funk. Why me? People were telling me to wait as long as possible before getting the surgery. Then, one day, I went to the blood specialist who needed to run some tests. I was in the same facility with 10 little kids, all hooked up to machines, apparently undergoing chemotherapy treatments. These kids were fighting for their lives, and I imagined that some of them weren't going to make it. But you wouldn't know it to watch them. Many were coloring and playing with toys. They were laughing together and supporting each other. They were definitely not lamenting whatever the future had in store for them.

Instantly, something inside me changed. *Mark, you selfish son of a bitch. How dare you say, "why me?" Who do you think you are that you get to live in a world without pain and fear?* These little kids were going through hell, but they weren't complaining. That's when I realized, if those kids could do it, so could I. Instantly, I went from a place of victimhood to a place of empowerment. I went from asking "why me," to "why not me?" I wasn't special. I didn't get to live a life free from pain and suffering. Everyone goes through that. So, I knew I had to stop feeling sorry for myself, stop being a victim of my circumstances, and get on living my life. That meant undergoing surgery and moving on. And in case you missed it, going from "why me? to "why not me?" was a new screen. One that gave me the powerful insight required to take action.

It's so easy for us to feel like a victim, but the reality is that victimhood is just another screen. And it's one of the most destructive and disempowering screens there is, because it renders you inactive. It gives you an excuse to not try. It gives you a reason to come up short and fail. This can happen to anyone, and if you take the time to stop and look around at the world today, you have more reasons than ever to play the victim. Just since 2020 alone, we've had to deal with COVID and the lockdowns when we were watching death clocks on the news every night. Once the global pandemic ended, things didn't get much better for families because inflation drove prices and the cost of living sky high. There was all this talk of a recession and market volatility. And if you can still manage not to get sucked down by all of the negative external circumstances of the world, everyone has their own internal struggles that come in the form of work troubles, family conflict, and illness.

My father told me a story about a king who once sent hundreds of wise men out into the kingdom to gather all of the knowledge available in the world. They returned with enough books to fill a massive library. The king didn't have time to read that many books, so he called the wise men back and asked them to consolidate what they learned. The wise men were able to get everything to fit into 500 books. "Still too much," said the king. "Cut it down." The wise men consolidated the books in half, but it was still too much. On and on this went until the wise men returned with a single sentence. It read, "There ain't no free lunch."

Everybody has a tough road. Nobody gets a free pass. It's so easy to see life through the screen of victimhood and entitlement. We're naturally wired to become self-absorbed and feel self-pity. I still catch myself doing it and saying, "Why me?" We all go to this place from time to time, and we all will go there again, but with the right tools, training, and development, you can make sure that you go there less. And when you do find yourself going there, you'll be better prepared to interrupt the pattern and get out of there faster. You don't have to get stuck there.

This begins with self-awareness and recognizing the victim mentality when it occurs, so you can work yourself out of it. Sometimes my wife needs to help me snap out of it. And I'm grateful that she does because I know the screen of victimhood will prevent me from living an extraordinary life. It will prevent me from experiencing freedom, fulfillment and love. You can't achieve something that you don't believe to be possible. So, if you believe you are a victim or the system is stacked against you, what chance do you have? If you believe you are entitled to a reward that you didn't work hard to achieve, you have no motivation to work hard. Victimhood prevents you from learning from a negative situation, or seeing any opportunities or silver linings that appear when things don't go your way. I learned that sometimes our suffering can serve a purpose or help achieve a greater good.

A few years after my hip surgery, my daughter was diagnosed with a terrible case of asthma. When the doctors were going to treat her with prednisone, I was able to intercede and prevent that. She was only 11, and not fully grown. If she had developed osteonecrosis as I did and required a hip replacement, she could have been crippled for the rest of her life. That's when I realized that what I had endured saved my daughter from suffering a worse fate. It's that attitude and perspective that can be the difference between you living an extraordinary life or not. So many of life's outcomes can be traced back to perspective.

Dr. David Eagleman turned me onto a fascinating study about aging brains.[2] Nuns throughout the country participated in this study. In addition to completing physical and cognitive tests, they provided detailed records about how they spent their days. When the participants passed away, they agreed to donate their brains for study. Researchers were looking for links between age-related brain degeneration and cognitive performance. But the results were unexpected. Nearly one-third of the brains tested had full-blown Alzheimer's disease even though the cognitive tests conducted on the participants when alive showed no signs of the disease. The reason was because the nuns remained highly active. They had responsibilities and learned new skills to keep the brain active, thus preventing them from experiencing any symptoms. Even as the brain tissue degenerates, mental and physical activity help the brain develop new pathways, or what is called *cognitive reserve.*

The very act of challenging yourself and living a life of continuous learning and development can offset the cognitive symptoms of a disease as devastating as Alzheimer's. That's remarkable and shows what is possible when you put in the effort and take responsibility for your actions, rather than succumbing to the limitations of a bad situation or throwing in the towel without trying. The participants in this study would not have experienced these results had they been sitting at home feeling sorry for themselves, behaving like a victim. If the nuns accepted the fact that they had Alzheimer's and would no longer be able to remain active, then they wouldn't bother. That's why the opposite of victimhood is responsibility, and that's the screen you must see the world through if you hope to live an extraordinary life.

I'm not saying there aren't victims in the world. Many people are born into poverty. People have suffered from abuse and undergone tragedy that forever changed the trajectory of their lives. You shouldn't ignore the feelings from those situations, but you can't let them define you. The question isn't whether you have been victimized in the past. The question is whether you will let that stop you from achieving success in the future. No matter what you have experienced, the victim mentality keeps you trapped in the past and prevents you from living an extraordinary life because it renders you powerless. You don't even need to be a real victim to have the victim mentality. I know people with millions who still believe they are victims. They may have money in the bank, but they certainly don't experience the freedom, fulfillment, and love that comes from living an extraordinary life. And they will never experience that until they change their mindset.

No matter what your circumstances, you can choose not to be a victim. It comes down to changing your screen. You are 100% responsible for your success. There are actions you can take today to ensure you are doing everything in your power to achieve freedom, fulfilment, and love in your life. And if you can learn how to take that bad thing that happened to you, or that setback, and turn it into something that can help others, you have truly embraced responsibility. We are all capable of reaching that level. It begins by being open to change.

Challenge Your Beliefs

To make true progress that will change your life, you need to do battle inside that last slice of the pie—the one that represents things you think are true but really are not. However, this transformation doesn't happen right away. To truly embrace new ideas and see the world through a new screen, you must first eliminate the old screen.

That can't happen if you don't admit that your old screens have a negative impact. If I show you the solution without you admitting that you have a problem, you will just continue with your old behaviors or drift back into that comfortable way of thinking and being when things get tough. This might require swallowing your pride and letting go of your ego. If you want an extraordinary life of freedom, fulfillment, and love, then you must first obliterate those disempowering screens and any beliefs that are factually untrue. Failing to do so is how people become their own worst enemy; it's how they remain trapped in an ordinary life. The problem is that succumbing to cognitive dissonance comes so naturally to us as human beings. We all want to look good, act smart, avoid being wrong, even when that means embracing self-pity and playing the victim. Although it may be natural, that is the surest way to not grow and learn.

I don't expect you to recognize everything right away, but once you start thinking like this and actively look for these inconsistencies, you will see them more clearly. Once you can identify the things you think are true that really are not, the next step is to make distinctions between your current screen and the new screen you hope to create, or the world in which you currently live and the new world of possibilities. Those distinctions can be about money, wealth, the nature of being an investor, how your brain works, or your blind spots. The clearer the distinctions, the easier it will be to loosen your grip on the old world and way of being. By making distinctions, you develop self-awareness and a new perspective that helps you better understand your options. That gives you power over your circumstances. If you don't make those distinctions, you might never realize that you have a choice or that you can take a different path. So, if you see the world through the screen of victimhood, entitlement, and expecting something for nothing, you can begin to distinguish that from the screen of responsibility.

Being able to recognize these distinctions with money and investing are especially important because popular misconceptions are dangerous. In Part II, you can learn certain academic principles that indicate common practices such as market timing, stock picking, and track record investing are not effective long-term investing strategies. As long as you're willing to accept it and not dismiss it because of cognitive dissonance, the logic will be waterproof and airtight. Once you can see that for yourself, choosing to create a new way of investing will be much easier.

If you can do that, it's a very good sign that you are capable of what my father used to consider a powerful weapon—being coachable.

That's the ability to take direction from someone else who has already achieved or knows how to achieve what you want in life. Not everybody is willing to seek out another person for guidance. As you'll learn in Part III when I explain why it's essential you work with a coach, nobody knows the value of coaching more than high performers, whether it's athletes, businesspeople, or investors. However, too many people feel threatened and think of getting help from others as a blow to their ego.

I bring this up now because I want you to be able to recognize the warning signs of being uncoachable when they appear. There will be times when reading this book when you will come to a fork in the road where you have two options—you can take the easy path that is familiar because it's the way you've always done things, or you can accept the coaching and do the hard work. It's by taking the second road where the majority of breakthrough thinking occurs that can result in an extraordinary life. Your dreams and creativity seldom exist on the first path. It will require perseverance and grit to overcome the challenges you will face along the way. It won't always be easy, but I promise that it will be worth it.

When I visited London, I had the opportunity to see Winston Churchill's underground bunker where he resided during parts of World War II. Churchill and the other leaders confined to those tiny quarters were under constant threat. One bomb could bury them alive. There was no assurance that they would win the war, yet despite the terrible conditions, they persevered. After being there inside that bunker, I walked away feeling a deep sense of gratitude for all of those who fought for the Allies against the Axis of Evil. That's the kind of grit and determination that it will take to live an extraordinary life. But remember, nothing worth doing in this life is easy. We do it because it's hard, not because it's easy. And living an extraordinary life will not be easy.

As you ponder the question posed at the beginning of the chapter and recognize more of your screens, take the next step and consider where that screen came from. How did you come to believe certain things? It might be friends, family, television, or an experience that led you to certain beliefs. This can help provide some much-needed perspective by helping you better understand why you think the way you do.

The Quest for Knowledge

Some people live their entire lives without doing any breakthrough thinking. But if you can harness this power, you are in good company. I've relied on several breakthrough thinkers to help develop the content and curriculum in this book. I've been extremely fortunate to work with experts such as Steve Zaffron and Dr. Terrance Odean, Dr. David Eagleman, and Nobel Prize winners Dr. Eugene Fama and Dr. Harry Markowitz.

You will discover things you don't know about yourself and how your brain functions when it comes to investing. But don't expect to know everything by the time you finish this book. This transformation requires a lifetime of incremental progress from a continuous pursuit of knowledge. That's how innovators like Walt Disney, Steve Jobs, and Elon Musk achieved success. It's how every great innovator has changed the world. They didn't focus only on what they knew they knew. They strived to learn what they didn't know they didn't know—things people never even thought possible. That required being unreasonable sometimes, but most significantly, it required asking and living with those questions for an extended period.

I am a big believer in knowledge. I love knowledge. There is so much out there to learn, and tremendous power comes with that knowledge. To steal a line from Nobel Prize–winning economist, and one of my mentors, Harry Markowitz, "I want to know what we know, and how we know it." That's why I've made reading and studying a priority in my life, and there is no better way to learn than through repetition.

I will warn you up front that you will experience repetition in this book. That's intentional. As laid out in the Figure 1.6, the first time you hear a new piece of information, you learn a lot, but you don't retain all of that information. The second time you hear

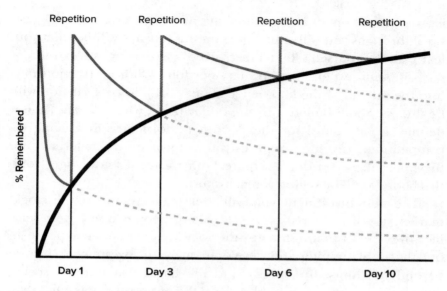

Figure 1.6   Spaced Repetition

Source: Steve Boller, "Spaced Learning and Repetition: How They Work and Why." Bottom Line Performance. October 16, 2012. www.bottomlineperformance .com/spaced-learning-and-repetition-why-they-work/.

that same piece of information, you retain a little more. The third time—you retain even more. This continues until the information sticks, and you have it down cold. Repetition is particularly important in this book because these concepts are connected. Each section will draw from and build on the information you learned in the previous section, so the better you understand these foundational concepts, the easier it will be to take action.

There are two very different but equally important types of knowledge—knowledge about yourself, and knowledge about everything else. Those are very broad categories because there is a lot going on in the universe that is not about you. Yet paradoxically, as human beings, we spend the majority of our time thinking about ourselves.

Human beings, as far as we can tell, are the only creatures on the planet who can think about how they think. That means we can teach ourselves about ourselves, and as far as I'm concerned, that is a miracle. It opens us up to making new discoveries. The biggest breakthroughs you will experience on this journey will be what you learn about yourself. Remember, it's about having an extraordinary life, not just an extraordinary portfolio. Without that journey into the self, all of the knowledge about portfolio construction will be useless. You'll be left with the booby prize, what *Merrian-Webster* defines as "an award for the poorest performance in a game of competition." Investing requires you to focus on the external and the internal. If you don't acquire both forms of knowledge from this book, you'll be selling yourself short.

The best investment you will ever make is not in the stock market, it's investing in yourself. Many people don't view self-improvement as an investment—they view it as work or an expense—but the right training and development is an investment. It goes beyond the hints, tips, tricks, and life hacks you find on social media. If it was that easy, we'd all have the answers by now. One reason why many people don't invest in themselves is that they can't yet see the greatness they possess. For the past 30 years, I've struggled to help people with this, but I've learned I can't do it for anyone else. Everyone must discover it for themselves.

Life is random and unpredictable. Bad things happen to good people. We can invest soundly and lose money. Circumstances will never be perfect, but I can promise that you will have the possibility to transform who you are in this world to live an extraordinary life. However, that's not the same as a guarantee.

When I coach people about expanding their business and investing their money, there are two excuses I hear all of the time about why they can't succeed. The first is "I don't have enough time." The second is "I don't have enough money." It's true that time and money are limited resources, but how you use them determines if you live an extraordinary life.

When I talk about living an extraordinary life, it's not hyperbole. *Extraordinary* is not a random word I plucked out of the sky. It is very intentional. An ordinary life is standard, typical, average, or expected. It's routine conformity, and there is nothing wrong with that. Lots of people are happy living ordinary lives, and if that works for them, more power to them. But if you want more, living an ordinary life is tragic.

An extraordinary life is not commonplace or standard. It's going beyond what is average or expected. It's about standing out by achieving great things and having a positive impact on others. It's living a life of freedom, fulfillment, and love. However, being extraordinary can come with pressure and high expectations as you manage various challenges, so you can grow. It's not easy. You must learn to properly allocate your most precious resources (your time, money, and also your energy) to what is most important in your life.

The responsibility will fall on your shoulders. You must do the work, because you are responsible for whatever state your life is in and you are responsible if you want to take it to the next level. You will have to battle some of your demons and dismantle previously existing screens. But you are stronger than you think. That work starts right now by defining your terms and creating one of the most powerful screens possible—a screen that enables you to live a life of freedom, fulfillment, and love, while embracing what is most important in your life. That screen is the American Dream.

Notes

1. Steve Zaffron and David Logan, *The Three Laws of Performance: Rewriting the Future of Your Organization and Your Life* (Jossey-Bass, 2009).
2. "The Brain with David Eagleman," Episode 2, *Nuns and the Aging Brain*, PBS, October 21, 2015. https://www.pbs.org/video/brain-david-eagleman-episode-2-clip-6/.

PART

I

Establish What's Most Important to You

"Nothing ever comes to one that is worth having except as a result of hard work."

—*Booker T. Washington*

The world is an amazing place, and if you help other people, you are rewarded for the value you create. That's been my father's attitude ever since he was a kid.

At 10 years old, my dad would get up early in the morning, pick up a stack of newspapers, and sell them on the streets of Charleston, West Virginia. When he ran out of papers to sell, he'd get out his shine box to shine shoes for money. Always on the lookout for the next opportunity, he read in a comic book about how Cloverine Salve can be used to help soothe cracked and dry skin. So, he went into business selling it to coal miners to protect their hands. He made sure to track down the miners on Fridays, right after they got their checks, and before they had a chance to spend it all at the bar. Before my father was a teenager, he was an entrepreneur three times over. That attitude and work ethic followed him into adulthood when starting his own financial advisory business.

Because my father developed a passion for the American Dream, serving others, and helping them invest and plan for their future, it was natural for me to follow in his footsteps. While other kids my age were reading comic books, I read books like *Think and Grow Rich* by Napoleon Hill. I was 12 years old when my father handed that book to me, and said, "Mark, you can learn anything you want to know by reading a book. Find someone who already has the results in life that you want to achieve and study their mindsets and behaviors. As you read, eliminate your own excuses and judgments about the strategies in the book. Do what it says to do, because take it from me—most people won't. It is all too easy for your inner voice to tell you it will never work or that it worked for them but can't work for you. You live in a great country, one that gives you the opportunity to build whatever you want in life if you're willing to work hard enough and create value for others."

He told me to reread that book every year if I wanted to fulfill my dreams. That's precisely what I did. I also made sure to put in the work. Every summer, I would push my lawnmower down the street. If I saw that someone's lawn was overgrown, I'd go knock on their door and offer to mow it. When it snowed in the winter and all of the other kids grabbed their sleds, I grabbed my shovel and went knocking on doors. Sometimes I'd get paid. Other times I'd work for hot chocolate and cookies. Forget the money, it was that feeling of pride and confidence that came from a job well done that was most rewarding.

My parents helped me understand that hard work is a virtue with tremendous payoffs. That's what the American Dream is all about. You can't experience the "fulfillment" in freedom, fulfilment, and love if you don't ever experience the benefits of hard work. And if you can improve the lives of others and help them succeed through your hard work, it's exponentially more fulfilling. That's what those who have the screen of victimhood will never understand.

Ideas are powerful, and I am so grateful that my father instilled in me the screen of the American Dream at a very young age. "America is not perfect," he would tell me. "But it's the best country on the face of the earth." If I hadn't had that screen at a young age and didn't believe that America was a great country where I could create anything I wanted, a book like *Think and Grown Rich* would not have resonated with me. If I thought like my grandfather, I would have believed that book was garbage. Instead, I read it and was blown away. It helped shape my vision of creating freedom and prosperity for myself and others.

That screen of the American Dream laid the groundwork for the actions I would take later in life—actions that others who didn't have that screen were unwilling to take because they didn't believe in the ideals and principles of the American Dream. It's why my obsession with learning continued as I grew older. By the time I got to college I devoured every economics, accounting, and finance book I could get my hands on. I crammed my class schedule with business courses. Even then, I sensed that to be successful and to live a powerful and extraordinary life, it would take more than an understanding of finance. It would mean studying human nature and how the mind works. The world seemed to be full of people with a massive amount of knowledge and intellect that were not generating the type of success I sought. There was more to achieving an extraordinary life than having an analytical view of business, and I was determined to find out exactly what that involved.

So, every summer during college, I went to work for my father. From the beginning, he made it clear that he wasn't going to hand me anything, and a full-time position with him was not guaranteed. He said, "If you work hard, go to college, and get a degree, you can come work at my office. If you do a good job, I'll keep you on. If not, I'll fire you immediately." There were no free rides, but that is also part of the American Dream. You need to take responsibility for your success. You need to earn it.

There was no better person to learn from than my father. A technically gifted guy, my dad spent a ton of time educating himself about financial planning. That's what made him such a phenomenal teacher. And I was a sponge, absorbing every word of instruction. I learned about financial planning, insurance, real estate deals, partnerships, mutual funds, annuities, and all the financial products at our disposal in the 1980s. I learned how to acquire, market, and sell to clients. By the time I graduated with a degree in accounting and finance in 1986, I already had my life and health insurance license, my fire and casualty license, and my stockbroker's license.

While my father gravitated more toward estate planning and the tax side of the business, I was fascinated by the investing side, which quickly became my focus. I realized that I wanted to become an entrepreneur, not just someone with a high-paying job I couldn't escape. I wanted to create a business that would grow and thrive, even in my absence, like Walt Disney, Henry Ford, and Thomas Edison before me. I believed I could do that as a financial planner by helping people invest their money and maximize their returns. Helping others achieve financial freedom so they could live an extraordinary life became my American Dream.

2

Defining the American Dream

"There are those who will say that the liberation of humanity, the freedom of man and mind, is nothing but a dream. They are right. It is the American Dream."

—Archibald MacLeish

"How do you know when you are experiencing the American Dream?"

Sometimes, it's best to define something by explaining what it's not. And what the American Dream is not about is money. Nor is it about your portfolio or the size of your house, because it has nothing to do with material possessions. The American Dream is also not something or someplace you get to one day in the future. "When I finally catch that big break and retire, I can experience the American Dream." The American Dream is definitely not about ill-gotten money. It can't be something that is taken from others by force or subterfuge. Unearned financial resources are destructive. The American Dream doesn't work like that because it is about something much bigger than money.

Simply put, I believe the American Dream is a screen through which you see the world. And when seeing the world through this screen, you have the freedom to live in a world driven by purpose so you and your family can experience fulfillment and love. It's an expression of creativity, and as far as I'm concerned, it's an expression of being alive. It's a commitment to and communication with others. It's self-expression and freedom from restraint. That means you create your American Dream through your communication, not your bank account. We live our entire lives through communication with all the people we speak to and meet with daily. I'm doing it right now by writing this book because I believe that the opportunity for others to live their dream is alive and well.

The *Oxford Dictionary* defines the American Dream as "the ideal that every citizen of the United States should have an equal opportunity to achieve success and prosperity through hard work." The term *American Dream* first appeared in the 1931 book *The Epic of America* by James Turslow Adams. He said, "The dream of a land in which life should be better and fuller for everyone with opportunity for each according to ability or achievement."[1] That can mean different things to different people because you determine what makes you happy and what your life is about. But no matter what you hope to achieve, the American Dream is rooted in freedom.

Our country may be politically divided, but our greatest leaders can agree on the importance of freedom. John F. Kennedy and Ronald Reagan, presidents from different sides of the political aisle, spoke passionately about freedom. But the concept of freedom predates all of our modern leaders. It is as old as the country itself, and was codified in the Declaration of Independence, the Constitution, and the Bill of Rights.

These three documents ensure Americans certain freedoms—the freedom of speech, freedom of the press, freedom of religion, and the freedom for the people to assemble peacefully and address their grievances. The founding fathers understood that private property was the foundation of prosperity and freedom, so these documents provide us with the freedom to use, acquire, and dispose of property. We have the freedom to own our own businesses.

In the United States, we also have the freedom to "life, liberty, and the pursuit of happiness." Happiness is defined as "the feeling of showing pleasure, contentment, well-being, or joy." Few countries mention the word *happiness* in their founding documents, never mind have it serve as the foundation for their principles and ideology. That is unique; it is part of what makes us lucky to live in this country. In the United States, you get to determine what makes you happy and what your life is about. You get to determine your relationships and how you function in the world.

What's also great about this country is that we have clearly defined laws and legal principles that apply equally to everyone to protect these rights, so we are not at the whims of the powerful, the violent, or the mob. These documents restrict the three branches of government from taking away our rights by limiting government power. Without these basic rights and freedoms, nobody can create a business. Without these basic rights and freedoms, you have oligarchies like you find in China or Russia.

These founding documents are what make the American Dream possible for everyone. It doesn't matter what your race, gender, religion, ethnicity, or political affiliation is; you are entitled to the American Dream. That is a universal principle for me. But despite the freedoms and opportunities available in this country today, the American Dream is under attack.

When I talk about the American Dream, people typically fall into one of three camps. The first really get it and believe in it to their core. My dad belonged to that camp. The American Dream was built into his DNA. Most entrepreneurs and small business owners feel the same way because they already reap the benefits of the American Dream. They don't need to be convinced about the importance of freedom and the opportunities at their feet. The second camp can acknowledge the possibility of the American Dream and want to achieve it, but struggle because of their circumstances. They believe it in theory, but don't know if it is possible for them. The final camp are the American Dream critics. Not only has this group given up on achieving the American Dream, they go so far as to think the Dream itself is bad, or evil. They believe the country is founded on greed, so they view the American Dream as the quest for materialism and think the world would be better off if people didn't have this idea in their heads.

I often hear these critics make numerous claims, such as housing no longer being affordable. They claim that too many people lack the resources to live comfortably in retirement. They claim that the middle class has contracted over the past five decades. At one of my workshops, I was confronted by someone who told me the American Dream was racist and a product of white privilege. Many of the critics ultimately believe that some people can't make it on their own; they are victims of their circumstances and can't improve without help from the government. They don't want the American Dream to fade away; they want to destroy it.

I'm certainly not saying that America is perfect. It's not. Times are tough, and people struggle. But that doesn't change the fact that the American Dream is alive and well. And I believe it is possible to achieve the American Dream no matter your circumstances or where you are in life. Here's how I know that:

- "43 percent of Americans planned to start a business in 2022."[2] That's incredibly high and pro-American. Try starting a business in Russia or Cuba and see how that goes.
- "Nearly 45 percent of Fortune 500 companies were founded by immigrants or their children."[3] That's pro-immigration. America continues to be the great melting pot.
- "Most Innovation Economy leaders (94%) expect revenue and sales to remain the same or increase in 2024. Similarly, 95% expect steady or increasing profits."[4]
- "The vast majority—81 percent—of today's young Americans feel that if they work hard, they will be able to succeed."[5] That's pro-American for sure.
- In 2023, small business employees made up 45.9% of the total number of US workforce.[6] The critics try to say that small businesses can't survive, and you must work for a giant Fortune 500 company to get ahead, but that's not true. Half of the country is employed by small businesses. That's because of entrepreneurship, innovation, hard work, and grit.
- People talk about a lack of opportunity, and complain about the economy, but it's never as bad as the critics make

it out to be. In January 2024, the unemployment rate was 3.7%—that's 6.1 million people.[7] But it's not like there weren't available jobs. At the end of December 2023, there were 9 million job openings.[8] That's almost 3 million more jobs than people unemployed, so there was work available for those willing to take it. Compare the current unemployment numbers to the height of the Great Depression in 1933 when unemployment was 24.9%.[9] Of course, conditions can always be better, but perspective is important.

I'll take the creators, those who believe in and live the American Dream, over the complainers every day, because the creators pave the way for the future. That requires taking 100% responsibility. Remember, the opposite of responsibility is victimhood. Over the years, I firmly came to believe this was my Grandpa George's screen, So, even when presented with opportunities, he turned them down because they didn't fit into his screen. Like trying to fit a round peg in a square hole, those opportunities didn't compute for him, so the idea of becoming an entrepreneur was unthinkable. That's why victimhood is one of the most destructive screens you can have as an individual. If you think that the world owes you something that you didn't earn, or that you can't make it on your own without help from the government, you will never be able to succeed with that mindset. Unfortunately, some people live their entire lives believing that they are victims. That belief can prevent people from achieving their potential and capitalizing on the opportunities available to them. However, none of these accomplishments would be possible if we did not live in a free country.

Out of 176 countries, the United States ranks 25th as the "most free," according to The Heritage Foundation (see Table 2.1).[10]

As a country, I believe we can do better and get higher up on this list—even up to number one. But if you look at the countries at the top (Singapore, Switzerland, Ireland, Taiwan, and New Zealand), these are much smaller countries. The United States remains the most prosperous large country in the world. Consider Figure 2.1.

Table 2.1   2023 Index of Economic Freedom Rank (1–176)

Least Free	Most Free
125 Russia	1 Singapore
131 India	2 Switzerland
154 China	3 Ireland
169 Iran	4 Taiwan
	5 New Zealand
	25 United States

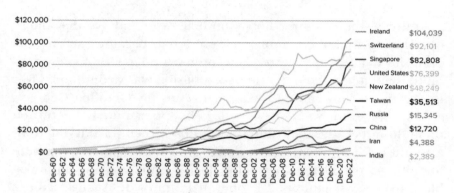

Figure 2.1   GDP per Capita Current US$ (1960–2022)

Note: GDP per capita is gross domestic product divided by midyear population. GDP is the sum of gross value added by all resident producers in the economy plus any product taxes and minus any subsidies not included in the value of the products. It is calculated without making deductions for depreciation of fabricated assets or for depletion and degradation of natural resources. Data are in current US dollars.

Source: World Bank national accounts data and OECD National Accounts data. Retrieved July 12, 2023. fileshttps://data.worldbank.org/indicator/NY.GDP.PCAP.CD?locations& name_desc=true. Taiwan Province of China GDP data retrieve March 22, 2023, from https://www.imf.org/external/datamapper/NGDPDPC@WEO/TWN.

We can still do better, but compare those countries at the top of the list with those at the bottom (Russia, India, China, and Iran) and you'll notice that the free countries all have a much higher GDP per capita. That can't help but influence quality of life, because with freedom and high GDP per capita comes satisfaction and the ability to live the life of your own creation.

In the book *Conscious Capitalism,* authors John Mackey and Rajendra Sisodia write, "Contrary to popular belief, prosperous countries also have a higher level of satisfaction. The self-determination associated with free markets and greater prosperity leads to greater happiness. The top quartile of economically free countries has a life satisfaction index of 7.5 out of 10 compared to the bottom quartile, which has an average life satisfaction index of 4.7."[11] The reason for this is that free countries provide the opportunity for people to be more prosperous, creative, enjoy better health, and live longer lives. But this freedom is not free.

Something else my father taught me when I was young was that freedom comes at a great price, and that price is measured in both treasure and blood. We've been attacked by outsiders, and the American way of life has been threatened. Our soldiers and military personnel over the centuries have paid a great price, but they aren't the only ones who have sacrificed for our freedom. During times of crisis, we have learned how critical our first responders are to ensuring our freedom. We all owe deep gratitude to those who exhibit bravery and courage when our country is threatened; we wouldn't be here and wouldn't have the opportunities we do without such heroes. Terrorists hate our freedom and want to destroy it. That's not going to change, which is why we need to defend ourselves. The American Dream depends on it. We owe a debt to the brave men and women who have kept this country safe and free. That is a debt we can never fully repay.

The founding documents of this country ensure our freedom as Americans. But I don't believe the ideals of the American Dream are restricted to only Americans. Arnold Schwarzenegger is the perfect example. I had the privilege of interviewing Arnold when he appeared at my company's Advanced Advisor Conference in 2024. Raised in socialist Austria, Arnold embraced the American Dream when he moved to the United States in 1968 at the age of 21 to pursue his dream of bodybuilding. He had lived through the failures of socialism and knew firsthand that the government wasn't the solution to society's problems. Nothing was handed to him and nothing he accomplished was easy. While training for competitions, he worked in construction and put himself through Santa Monica City College. During my interview with Arnold, he spoke about the importance of hard work and many of the same virtues discussed in this book. Those beliefs remained with him throughout his bodybuilding and acting career, but it wasn't only about him. He wanted to give back, which is why he went into politics and was elected governor of California in 2003. During his tenure, the state experienced tremendous growth because Arnold firmly believed that it was the government's job to help the private sector instead of derailing it. He was pro-entrepreneur and pro-capitalism, and because of that, what he accomplished was not possible in many other countries.

My good friend Gary Sinise shared this about his American Dream:

> *My first job working for someone other than my father doing yard work was at 13 years old helping out a the local pizzeria. I enjoyed the little paycheck, but it did not last long as I sluffed off and was let go. No more paycheck. A hard lesson, but a good one. Work hard and there are rewards; sluff off and "see you later, you're outta here." So, at an early age I became a do-it-yourself, work-hard kinda kid. As a cofounder of Steppenwolf Theatre in Chicago at 18 years old, these work hard lessons paid off, as what we created as kids all those years ago is now a world-renown 50-year-old Chicago institution. It is truly a great American Dream story, and example of passion, commitment, perseverance, and drive all coming together and how the freedoms we enjoy here in America allow us to "dream big," succeed financially, and achieve what we set our hearts to.*

Gary nails the essence of the American Dream in this story.

I want to see the principles and ideals of the American Dream spread worldwide. I believe the American Dream can be a global phenomenon. Freedom, self-determination, and opportunity will make prosperity and growth available everywhere. But just as Arnold very much understood, that can only happen with the right economic system in place.

The Power of Capitalism and Free Markets

If freedom is the foundation of the American Dream, the engine that makes that dream a reality is capitalism and the free market economy, because that is what drives wealth creation.

Capitalism is "an economic system in which private actors own and control property in accord with their interests, and demand and supply freely set prices in markets in a way that can serve the best interests of society."[12] If you believe in freedom for all mankind, creativity, peace, prosperity, self-determination, accomplishment, free markets, innovation, upward mobility, love, entrepreneurship, private property, giving to others, and creating value, congratulations! You believe in capitalism.

Capitalism is based on the ideals of the American Dream; I believe it is the only economic system that enables one to realize the American Dream, because it inspires competition, innovation, and wealth creation. I also believe that capitalism is the most moral economic system because it's based on the consumer. The consumer, not the government, determines who has wealth in capitalism. With capitalism, consumers "vote" by spending money on products and services. Not all economic systems enable these same opportunities.

Socialism and communism, for example, involve significant government control. The government sets pricing, as opposed to the rule of supply and demand or competition, as in a free market system. Government officials have the power to confiscate wealth from the people who created it and distribute it to those who didn't earn it. In communist countries, the government also owns the means of production. Not only do they set prices and confiscate wealth but they also take over the entire apparatus to become the means of production. That doesn't allow room for corporations or the ownership of private property, which is what makes competition and capitalism possible.

I believe Ayn Rand described the perils of socialism best when saying, "If a businessman makes a mistake, he suffers the consequences. If a bureaucrat makes a mistake, you suffer the consequences."[13]

John Mackey and Rajendra Sisodia argue in *Conscious Capitalism* that "in the long arc of history, no human creation has had a greater positive impact on more people more rapidly than free enterprise and capitalism."[14] It's not only the 1% who benefited, but billions of people worldwide. Don't believe me? Two hundred years ago, 85% of the world's population lived in extreme poverty (living on less than $1 a day). Today, that number is only 16%.[15] Yes, this number should be zero. Nobody should have to live in poverty. But there is no denying that poverty has significantly declined with the spread of capitalism. It has created prosperity for billions of people around the globe. There is no better example of upward mobility than the United States (see Figure 2.2).

According to Figure 2.2, in 1967, 24% of Americans lived in poverty as measured by household income. Today, only 16% of Americans live in poverty ($15,000–$24,000). The lower middle class ($25,000–$50,000) is becoming the middle class ($50,000–$90,000), the middle class is becoming the upper middle class

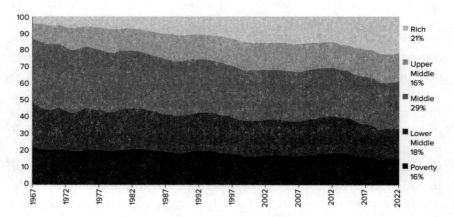

Figure 2.2   Share of the US Population in Each Income Class (1967–2022)
Source: US Census Bureau, Households by Total Money Income, Race, and Hispanic Origin of Householder: 1967 to 2022. Retrieved October 5, 2023, from https://www.census.gov/library/publications/2023/demo/p60-279.html. Census.gov defines poverty as total money income under $15,000 to $24,999, lower middle $25,000 to $49,999, middle $50,000 to $99,999, upper middle $100,000 to $149,999, and rich $150,000 and over.

($100,000–$149,000), and the upper middle class is becoming increasingly wealthy (over $150,000).

We want more people in the upper middle class, and we definitely want fewer people in poverty. The critics don't address these facts because they don't line up with the screen through which they see the world. Yes, the rich are getting richer, and more people are becoming wealthy, but the idea that the middle class is getting poorer is simply false.

Still, many believe that capitalism is a system based on greed and motivated by self-interest. Even Adam Smith, the father of capitalism, believed that people would compete because they only wanted to benefit themselves. I'm not denying that self-interest exists and that some capitalists are driven by greed. I know that is true. However, I also know the vast majority of entrepreneurs are driven by the desire to create products and services to help other people. They have altruistic goals to make the world a better place. For them, it's an act of creativity.

Entrepreneurs identify a need and then create products to fill that need. Next-level entrepreneurs like Steve Jobs, Jeff Bezos, and Mark Zuckerberg took this a step further by creating products that people didn't yet realize they wanted or needed—products people didn't even believe possible only to learn they couldn't live without them. The fax machine, the ATM, and the smartphone were all products I didn't think I would have any use for when they were first introduced, but they all quickly became essential at the time.

That act of creation and making the world a better place is at the heart of the American Dream, and it is not a selfish act. A socialist asks, "How can I confiscate money that's been created by others and distribute it to myself and my power base underneath me?" A capitalist asks, "How can I create value for others?" That's what motivates capitalists and that's what the system rewards.

It's funny how people believe capitalists are greedy but give politicians a free pass. The only thing more seductive than money is power, and the politicians who take freedom from the people do so because power is seductive. One of the biggest economic lies told by the opponents of capitalism is that those who create wealth and prosperity are so greedy they hoard their money, so it never benefits society or others. This assertion is blatantly false. There are multiple things those with wealth can do with their money.

First, people can spend it. Whether they buy a house, a car, a boat, food, or clothes, that money goes back into the economy to create growth, jobs, and opportunity for others.

Second, people can invest that money. Investing in stocks, bonds, equities, real estate, or their own companies also grows the economy. When you invest, you can make the world a better place by providing capital to fuel the engine that works to increase the standard of living for billions of people on the planet.

Finally, they can give it away to charities or other sources. Capitalism is amazing and directly responsible for global prosperity over the past 200 years, but it's not perfect and leaves gaps. Some needs are not addressed by pure capitalism. And where there are gaps, pure capitalists tend to be extremely generous. Americans donated an estimated $485 billion to charities in 2021.[16] That's more money than any other country. In other words, America is the most generous country in the world. You don't hear that often on cable news. When you're prosperous and work hard, you have the capital to help people in need. Living the American Dream also means being of service to others, and capitalism is what makes that possible. That's how you know you have an economic system that works.

Contrary to what some people believe, wealthy individuals don't hide their money in a vault, and they don't bury it in their backyard. In other words, they don't hoard it. That's not how the wealthiest people think and behave. They constantly use it to grow the economy and the quality of life on the planet. But no matter where the wealthy put their money, it's being reinvested somewhere.

That said, free markets and capitalism create equal opportunity, not equal outcomes. Those who believe in equal outcomes feel that it's unfair that some people have so much and others so little. They seek to level the playing field so everyone has the same. The problem is that's not how our system was designed by the founding fathers. However, even those who say they want equal outcomes don't really understand what that truly means.

First, it would mean that money was evenly distributed to everyone in the world. To do that, we would have to take the $454 trillion in global wealth, subtract the $305 trillion in global debt, and then distribute the remaining $149 trillion to the 8.1 billion people on the planet. That would leave every individual with $18,395. What kind of impact would that have? Would it make the world a better place? No, of course not. It would destroy the global economy and send our species back to the stone age. Then what would happen? Unless you continue to suppress freedom and confiscate wealth, over time, those who work hard to create value will be rewarded with more money and those who don't will have less money. Ultimately, we'll be right back on the path to where we are today. It will never work.

In reality, nobody thinks true equal outcomes are a good idea. None of the politicians who promote this policy are suggesting they hand over all of their own money, so they can live off $18,395 like everyone else. What they really want to do is confiscate money from those who are not in their "preferred" group and distribute it to those who are, be it their friends, voters, constituents, those on their boards, and those who pay them consulting fees and for speaking engagements.

Besides being completely impractical, the bigger problem with the concept of equal outcomes is that it's a by-product of the victim mentality. It would destroy the incentive to work and reward failure by confiscating money from those who earned it and distributing it to those who did not. Incentives matter, and you get more of what you reward. Those who receive money or wealth they did not create or earn will never get the reward that comes from creativity, self-expression, and hard work, because being handed money or an outcome for something you didn't earn will not make you any happier.

You can have either equal outcomes or equal opportunity—you can't have both. Equal outcomes pull everyone down to the same low bar. But freedom enables people to create value and earn capital. It's only through that capital that we create so many amazing technological breakthroughs. We couldn't do that if every person had only $18,395. It's the capital that creates wealth, prosperity, and improves the quality of life, not equal outcomes. Equal outcomes require suppression and control. It takes freedom away from the people, and it's the complete opposite of the American Dream. Destroy incentives and you destroy capitalism. Destroy capitalism and you destroy the American Dream.

In a capitalist society, inevitably, there will be people who create more value for consumers and society than others. No one is entitled to success, which means there will be winners and losers. If someone comes up with a better smartphone, they will capture the market share. In a free market system, the consumer, not the government, determines who benefits. If enough people want your product, you will sell more of that product than your competitors. If people don't want your product, you will not be as successful. Even if your product was popular for 100 years, as soon as it's no longer considered useful, you evolve out of the system. That's what keeps capitalists honest.

You may not think Bill Gates should have so much wealth, but the market obviously does because, in a capitalist system, you are rewarded for creating value. As my father liked to remind me, until you provide value, nobody owes you anything, but with hard work and determination, you can create anything you want. That is part of the American Dream and one of the many reasons why it's an extremely powerful screen through which you see the world. For those who have it, certain actions become obvious because your screen determines your actions. If I didn't have that screen, I wouldn't have created my business and capitalized on opportunities when they presented themselves. And as a modern-day American citizen, you have more opportunities available today than ever.

The Greatest Time to Be Alive

Some people ask how I can talk about achieving freedom, fulfillment, and love at a time when the circumstances for some people are so grim. It's true that the playing field may not be equal for everyone. It's not, but it sure didn't prevent my parents from escaping poverty, nor did it stop the millions of others who came from nothing to create wealth, prosperity, and live the American Dream. Yes, some people have more resources at their disposal than others, and far too many people still live in poverty, but for those who believe that circumstances make it impossible for them to succeed or that America is immoral or has gone off the rails in some way, I have one question. *Compared to what?* In what other country can you start off in destitute poverty and not only move into the middle class, but become extremely wealthy? That is unique and what makes America so special.

No place or civilization on this earth is perfect. Critics of the American Dream must remember that there are places in this world today where genocide occurs. Dictators murder their political enemies and suppress their people. There is still actual slavery on this planet. Not 150 years ago. Today! In 2022, it was estimated that a total of 49.6 *million* people live in modern slavery, a quarter of whom are children.[17] Many countries around the world don't have adequate food and health care, so many struggle to survive, and have little hope for the future. It's easy to forget how many people around the world still die from curable and preventable diseases, such as tuberculosis (which claimed 1.3 million lives in 2022[18]) and malaria (which killed an estimated 608,000 people in 2022[19]). So, if you're waiting for all the ills of the world to be solved before you start living your American Dream, you're going to be waiting for a very long time.

America offers exceptional opportunities for people to experience a high standard of living and thrive. It may not be perfect, but it's arguably the best country to live in the world. But while living comfortably in America, it's easy to lose perspective and forget about the harsh realities that exist in other corners of the globe. There is a reason why nobody gets into a rickety boat and risks their life to sail from Miami to Cuba, the way many refugees come to America. Why do so many people want to live in this country and try to get here at all costs? Some people do come here to commit crime and exploit the system. Still, there are many people who risk their lives to come here because they believe in the American Dream. They want the experience that comes with being an American. They want to be rewarded for their hard work and the value they create. They want to come to America because they want to live an extraordinary life. They want to feel the pride, self-esteem, confidence, and fulfilment that comes with a job well done.

I would argue that not only is America one of the greatest countries to live on the face of the earth and has more opportunities available than ever before but also that this is the best time to be alive in the history of planet earth. The computer you carry around in your pocket that you call your phone is more technologically advanced than the instruments first used to send men to the moon. "Today's smart phones have more computer power than all of NASA did back in 1969."[20]

You live better today than the king of England did 200 years ago, when there was no electricity, indoor plumbing, running water, refrigeration, or motorized transportation. Don't get me started on medicine. They didn't have antibiotics 200 years ago, so if you got an infection, there was a good chance you might die. If you needed surgery, there was no anesthesia. There were no cataract surgeries or hip replacements, no bypasses or stents. Most people lived in absolute squalor. Modern medicine is nothing short of a miracle, as I know this firsthand; if it weren't for modern medicine, I would have been in a wheelchair after my osteonecrosis diagnosis. In our house, we sometimes play a game called, "who'd be dead?" We go around the table and list off all of the injuries and ailments that are considered minor by today's standards but could have easily killed us in the past. I know it's morbid, but it's an excellent reminder of how lucky we are to be born in this time.

Capitalism has helped pull billions of people out of poverty and led to the expansion of the global middle class. Life expectancy, before COVID, was on the rise, as is the world literacy rate. We are surrounded by miracle after miracle. We should all wake up every morning, jump out of bed, and shout about how happy we are to be alive. But do we? Of course not. We take those miracles for granted. Instead, we're much more likely to say, "Life is so hard." It's easier to be a critic and play the victim than to consider all the opportunities at our fingertips.

It all goes back to the screen through which you see the world. If you choose the screen of victimhood, it will forever limit the opportunity and possibilities that are available to you. Because if you don't believe something is possible, how can you ever expect to achieve it? Ideas are so powerful that they don't even need to be true. Look at my grandfather. Seeing the world through a specific screen will determine your actions and, eventually, your results. It's a self-fulfilling prophecy. So why not create a screen empowering you to achieve freedom, fulfillment, and love? That's what is possible when you have the screen of the American Dream.

Experiencing the American Dream

It's one thing to be able to define the American Dream and understand the benefits. It's something else entirely to experience and live the American Dream.

Think about the concept of love. You can read a sonnet about love, study marriage and relationships, learn what happens to the chemicals in the brain when one is in love, and watch all of the romcoms that have ever been made to accumulate a comprehensive knowledge about the concept of love, but that's not the same thing as experiencing it or being in love. It won't prepare you for the moment you walk down the aisle to get married or when you hold your child in your arms for the very first time. The concept and the experience of love are two very different things.

The American Dream works the same way. You can't hope to achieve it by watching others or reading a book. It can only be created through thought and action. And once you can experience it, dwell in it, take it out into the world, and share it with others, that opens you up to a whole new world of possibilities. It will completely change your life. And nobody can take it away from you. You can surrender it or give it up, but nobody can take it. In *The Old Man and the Sea*, one of my favorite authors, Ernest Hemingway, wrote, "A man can be destroyed but not defeated." In other words, nobody can take away your ideas, your honor, and your values. You can only be defeated by giving up or giving in.

I don't ever want to forget or take for granted the American Dream. That's why every day, I express my gratitude for the opportunity to live in America, where I have the freedom to choose and pursue my American Dream. And I'm grateful to my parents for instilling those values in me. You, too, have the freedom to create and experience the American Dream. It won't necessarily be easy, but through initiative, hard work, and determination, you can achieve what's most important to you in life. All of this sounds good on paper, of course, but let's think back on the question I posed at the beginning of the chapter: *How do you know when you are experiencing the American Dream?*

This is a question I've pondered for decades, and it's one I can't help you answer. Only you can define and live your own dream. You are 100% responsible for creating it. No one else will create it for you. No government. No business. No other person. You create it for yourself. But don't rush to answer that question, because this is not something you can force. I liken the American Dream to happiness, because the more you try to bring happiness into your life, the less happiness you have. That's because happiness is a by-product of something else, so striving to get more doesn't work. The same thing is true of the American Dream. Try to seek out the American Dream, and you only push it further away.

Instead, consider who lives inside your American Dream. Today, my wife Melissa is a major part of my American Dream. We were each previously married, and we each had three kids when we first met, so we promised each other that we'd not have any more kids. Or pets. But after we were married, Melissa started hinting that we should have more kids. "I love you so much," she said, "I can't imagine living the rest of our lives without children of our own."

Since then, we have had two more kids of our own and two dogs. We're now a family with eight kids and two dogs. Melissa would like to have 50 grandkids, and with eight kids, we might get close. I love having a big family. It keeps us busy, and it keeps us young. We want the same things that most people want. We want our kids to grow up healthy and to get a good education. We want to watch our grandkids grow up and be able to spend time with them.

As you can see, family is very important to me. If you're anything like me, your family lives inside your American Dream. However, that dream is not limited to only family. Think of friends and those who matter the most to you. Think of the community where you live and work. Think of all of the people you hope to influence and make a difference for, and then begin to help them achieve their dreams. The more you stop thinking about what you want as an individual and focus on helping others, the more your dream will come into focus. Start by helping those people who mean the most to you because that is by far the best way for you to experience the American Dream. It's okay if that doesn't make sense right now. Don't worry; it will.

I want you to experience the American Dream and reap all the benefits this great country has to offer. I want you to have the opportunity to create and innovate. That dream is within your grasp, and as we will discuss in detail, it's directly affected by your saving, planning, and investing strategies. So, it's crucial that you invest your money prudently and with purpose. However, it's important to remember that money is only a tool. It's not the dream itself. And like any tool, you must be trained to use it responsibly, because if you're not careful, it can do a tremendous amount of damage, as well.

Notes

1. James Turslow Adams, *The Epic of America* (Transaction Publishers, 1931).
2. Shawn Langlois, "Nearly Half of Americans Have Considered Starting Their Own Business." *Yahoo Finance*, May 17, 2024. https://finance.yahoo.com/news/nearly-half-americans-considered-starting-140058309.html.

3. "New Report Reveals Immigrant Roots of Fortune 500 Companies," American Immigration Council, August 29, 2023. https://www .americanimmigrationcouncil.org/news/new-report-reveals-immigrant-roots-fortune-500-companies#:~:text=The%20 report%2C%20%E2%80%9CNew%20American%20 Fortune,by%20immigrants%20or%20their%20children.

4. "2024 Business Leaders Outlook: Innovation Economy Around the World." JP Morgan, 2024. https://www.jpmorgan.com/ content/dam/jpmorgan/documents/cb/insights/outlook/ business-leaders-outlook/cb-2024-business-leaders-outlook-ie-international-ada.pdf.

5. "Opening Doors to Opportunity: Generation Z and Millennials Speak," October 2020 Research Report, Walton Family Foundation, Echelon Insights, 2020. https://8ce82b94a8c4fdc3 ea6d-b1d233e3bc3cb10858bea65ff05e18f2.ssl.cf2.rackcdn .com/b1/02/ddcbc1d6434d91e8494f0070fa96/echelon-insights-walton-family-foundation-generation-z-millennials-and-opportunity-report-october-2020-10-6-20.pdf.

6. "How Many Small Businesses Are There in the US in 2023?" Oberlo. https://www.oberlo.com/statistics/number-of-small-business-in-the-us.

7. "TED: The Economics Daily," US Bureau of Labor Statistics, February 8, 2024. https://www.bls.gov/opub/ted/2024/number-of-unemployed-at-6-1-million-in-january-2024.htm#:~:text=In% 20January%202024%2C%20the%20unemployment,6.1% 20million%20in%20January%202024.

8. "Job Openings and Labor Turnover Summary," US Bureau of Labor Statistics, January 30, 2024. https://www.bls.gov/news .release/jolts.nr0.htm.

9. "Great Depression Facts," FDR Library & Museum. https://www .fdrlibrary.org/great-depression-facts#:~:text=How%20high%20 was%20unemployment%20during,or%2012%2C830%2C000%20 people%20was%20unemployed.

10. Patrick Tyrrell and Anthony Kim, "2023 Index of Economic Freedom: Economic Freedom Declining Worldwide," The Heritage Foundation, February 2024. https://www.heritage.org/ index/pages/all-country-scores.

11. John Mackey and Raj Sisodia, *Conscious Capitalism* (Harvard Business Review Press, 2013), p. 14.
12. "What Is Capitalism?" *Finance & Development* 52, no. 2 (June 2015). https://www.imf.org/external/pubs/ft/fandd/2015/06/basics.htm.
13. Ayn Rand, *The Ayn Rand Letter* (Penguin, 1988).
14. Mackey and Sisodia, *Conscious Capitalism.*
15. Ibid., p. 12.
16. Giving USA. 2022. https://givingusa.org/wp-content/uploads/2022/06/GivingUSA2022_Infographic.pdf.
17. "What Is Modern Slavery," Anti-Slavery International, 2022. https://www.antislavery.org/slavery-today/modern-slavery/#:~:text=According%20to%20the%20latest%20Global,of%20modern%20slavery%20are%20children.
18. "Tuberculosis," World Health Organization, November 2023. https://www.who.int/news-room/fact-sheets/detail/tuberculosis.
19. Richard Linga, "World Malaria Report 2023: Key Findings from the Report," Target Malaria, December 4, 2023. https://targetmalaria.org/latest/news/world-malaria-report-2023-key-findings-from-the-report/#:~:text=Since%202000%2C%20malaria%20deaths%20declined,deaths%20than%20before%20the%20pandemic.
20. Wendy Glittleston, "Today's Computers vs. the Apollo 11 Moon Landing Machine," Galvanize, July 15, 2021. https://www.galvanize.com/blog/todays-computers-vs-the-apollo-11-moon-landing-machine/#:~:text=Today's%20cellphones%20have%20more%20computer,gossipy%20texts%20about%20his%20colleagues.

Transforming Your Relationship with Money

"Monsters are real, and ghosts are real, too. They live inside of us and sometimes, they win."

—*Stephen King*

"What is your relationship to money and investing, and how does that show up in your life?"

Stephen was an investor who grew up poor. His father was on disability and couldn't work, so his mom was the breadwinner. At any given time, she had three different jobs. Their circumstances were unfortunate, and that created a tremendous amount of tension around money.

Stephen was the youngest of four children. He was entering high school when his older sisters were preparing for college. He recalls this being a particularly stressful time for the family. Money was never discussed openly, but it was a source of anxiety, stress, and fear. He heard his parents argue about money. At one point, he wasn't allowed to answer the phone because bill collectors would often call the house. Over time, an attitude about money developed, and that instilled in Stephen the idea that only bad people made money. That meant because he viewed his family as good, it didn't matter how hard they worked; they would never earn enough, because he didn't want to "see them" as evil.

Now in his forties and with a family of his own, Stephen has been unable to shake that attitude about money he developed as a child, even though he makes a respectable salary. His situation is much more stable now than it was for his family growing up. But he still lives with a scarcity mindset regarding money. In his mind, money has a negative connotation, and he's always waiting for the other shoe to drop.

Although Stephen's circumstances are unique, his behaviors and attitudes about money are quite common. They contribute to his relationship with money, and the financial demons that he carried with him into adulthood.

The No-Talk Rule

When growing up, did your family talk about money?

If you're like most people, your family seldom discussed money in a meaningful way. You didn't ask how much money your parents or someone else made. You didn't ask how much certain things cost. It was considered rude or impolite. This "no-talk rule" wasn't decided on or even spoken about; it was simply implied. In other words, there was a no-talk rule about the no-talk rule.

I work with many people who believe they are open and honest with their family about money, but on closer examination, it becomes clear that there are certain things they don't discuss, or there is a line they don't cross. The reason is because most people have a negative relationship with money.

I've seen money tear families apart. It can create jealousy, resentment, anger, mistrust, and envy among loved ones. It can lead to feelings of stress, anxiety, fear, struggle, shame, guilt, embarrassment, and, in some cases, even pain. Money has perhaps done more damage to families than anything else I can think of. And since nobody likes to discuss topics that evoke all those negative feelings, we avoid it, or come up with excuses that prevent us from having to talk about money, by saying it's impolite to have those conversations. Typically, the things we don't talk about are the things that make us feel shameful and guilty. And you can't live an extraordinary life if you feel shameful and guilty about money.

There is a psychological component to the no-talk rule. As you'll learn more about in Chapter 6, the human brain wants to remember good experiences and discount, minimize, or forget the bad experiences. Why? Our brains want us to feel smart and powerful and grounded. That's why people "fake it until they make it" by acting like they have it all figured out. But it's an inauthentic way to live that often creates even more stress.

What's so interesting about money and its relationship to human behavior is that these feelings are true for both those who have money and those who don't. The wealthy aren't spared. Just because you have money doesn't mean you experience fulfillment or consider it empowering. Some of the most unhappy clients I've worked with, and those with the unhealthiest relationship with money, have the most.

On the surface, the no-talk rule seems harmless. But just because families may not directly speak about money, doesn't mean they aren't indirectly conveying a message to each other and their children that shapes their attitudes and ultimately their relationship with money. And more often than not, that relationship is destructive. When parents don't speak openly and honestly with their children about money, it prevents kids from learning sound financial saving and investing principles. And if you don't talk with your kids about money, or you draw a line when you reach certain topics, how will they ever learn? I've worked with investors who have admitted that they graduated from high school not knowing how to write a check or open a savings account. I've heard horror stories from people who came from affluent homes and faced insurmountable credit card debt once they were on their own.

If you grew up in a household with some version of the no-talk rule about money, did that rule carry over to adulthood and influence how you talk (or don't talk) with your family about money today? Ask yourself if you are willing to break that rule. If so, the first step is very simple: *talk about the no-talk rule.*

That's how you begin to destroy it. Simply share your experience with the no-talk rule with those who live inside your American Dream. If you take that one simple step, you'll discover you are on the right track to improving your relationship with money, while helping your family do the same.

No matter how strict the no-talk rule is inside the house, no such rule exists outside of the house. When it comes to television, news, music, social media, or popular culture in general, you can't help but be bombarded with opinions about money. And rarely are those sources reliable or looking out for your best interest. That conversation about money happening out there among friends, on cable news, and on the internet is disempowering, but that's how kids and young adults learn if they don't get these lessons at home. Society puts such a high value on money that it becomes easy to equate money with our status and self-worth. It can determine how people look at you or, more accurately, how you believe they look at you. It can also determine how you look at yourself.

All of these outside influences can create a profound misunderstanding and misinterpretation about money that can have disastrous effects on our financial well-being and lead you into a destructive tailspin of always wanting more. When left to fester, these desires can haunt us.

Identify Your Money Demons

We all want to pretend that we are rational when it comes to money. We want to believe that we never get stressed or scared about finances. Nobody wants to admit that they aren't in control. But I encourage you to be honest with yourself as you ponder this chapter's question: *What is your relationship to money and investing, and how does that show up in your life?*

Don't think about how you hope or wish you felt about money. Focus only on your current beliefs—the good, the bad, and the ugly.

I ask investors, employees, and those who attend my workshops this question. Here are some of their possible answers. Do you see any of your own concerns about money in their responses?

- Money? There is nothing that I'd like to talk about less.
- I wake up every morning and go to bed every night worried about money. Literally, every decision I make, and every waking moment is clouded by the idea of not having enough money. I fear running out of money sometime in the future, leading to a life of constant worry.
- My parents believed that money is evil and the people who have it are greedy.
- I'm scared I won't have enough money to leave my family after I'm gone.
- My relationship with money is not a good one. I would love the feeling of being free from worry and concern.
- My husband and I have been lucky to experience abundance, but I don't feel free. It's as if I'm not worthy to spend the money I have. Anytime I buy something that's even a little expensive, it always feels like I don't deserve it.
- I was always stressed when I didn't make enough money. As I got older and started making more, the stress didn't go away; it was just different because I was suddenly scared that I would lose it. Once I made more, I needed more, so I was damned if I do and damned if I don't.
- I've worked hard my entire life, but I just haven't been able to save the amount of money I know I should. It makes me feel like a failure.
- I'm okay now financially, but I'm worried I won't have enough money in an emergency.
- I've done well and made good money, but I'm still not where I want to be.
- I was raised to believe money is a zero-sum game: for someone to win, someone else must lose.
- Money: it's difficult to make and even harder to keep.

- I feel like I'm on a treadmill. I struggle to make money and pay my bills at the end of every month, only to start all over again at the beginning of the next month.
- I believed that I couldn't make money if I didn't hustle. It's negatively affected my health and robbed me of time that I could have spent with my family and doing what makes me happy.

As you can see, these aren't constructive beliefs or feelings. They don't give you peace of mind and the feeling of security and abundance that seems so elusive. In fact, they do the complete opposite, which is why I refer to them as *money demons*. Left unchecked, these demons can eat away at you from the inside. And it doesn't matter how much money you have. Anybody can become infected.

I've been working with families for more than 40 years. During that time, I've noticed patterns in how people think about money, and many people's money demons can be traced back to four common associations or feelings about money.

Survival

For many people, money is viewed as a source of survival. Part of this is true. To survive in life, we absolutely need money. But we don't need as much as we think, and it certainly shouldn't be something we stress over and fight to get. There is an absolute minimum of money we all need to survive, but I seriously doubt that anybody reading this book is at the level where they struggle to get the bare minimum. You likely have more than you need, but you're still fighting to get more. Ask yourself why that is.

Too many people believe their American Dream is simply reaching a place of financial security. I'm sorry, but that's just not acceptable to me. I want to help those hard-working people, because I believe that the American Dream is about so much more than financial survival.

This survival mindset is propelled by the idea that money is scarce, and there is only so much of it available. It leads people to compete to get more money as a survival mechanism. They feel they must protect themselves, because they believe if someone has more that means someone else has less. That couldn't be further from the truth. Money and the accumulation of wealth is not a zero-sum game. The decrease in global poverty discussed in Chapter 2 is proof of that. Yes, the rich are getting richer, but the poor are also getting richer. The reality is that money is abundant. Next time you're in an airplane, look down at the massive cities, residential areas, and farmland 35,000 feet below. It takes money to build and sustain all of that, and that money is available to anyone who is willing to work hard and create value for others.

Sometimes this survival mindset propels people to work harder. It's motivation, but it's not a positive relationship with money or a comfortable way to live. Especially if you cling to your money and never see how it can work for you and grow. Where's the peace of mind in that? It can't exist in that scenario.

Fear

You may look at the world today and see chaos. Times are tough, and the future is uncertain. I understand why many people are fearful about money, but the reality is that this is not anything new.

I've helped investors navigate and survive a series of crises. I worked with investors when the dot-com bubble burst in 2000, and people's futures were gutted. I remember when those two planes hit the Twin Towers on 9/11, flights across the US were grounded, and the economy seemed to come to a halt. I was there in 2008 and 2009 when it looked like the entire global economic system was about to implode due to the financial crisis.

There are also personal circumstances that can send anyone's life into a tailspin. A child can get sick, a loved one can pass away, or you could experience an unexpected and financially devastating medical issue. You could lose your job, or your 401(k) could tank and plummet in value by 50% or more. Numerous events can destroy your business, financial portfolio, and even your outlook on life.

It's human nature to panic and assume that the current crisis is the worst. However, we've been through tough times before, and I can promise you that there is a way through. That's important, because even though the chaos we're living through will pass, there will be more chaos, fear, and uncertainty in the future. It means that uncertainty is constant, so we must learn to focus on the positive. I frequently remind people in my workshops that tough times didn't come to stay, they come to pass.

Evil

My grandfather believed money was evil, so only greedy and bad people had money, but he certainly wasn't the only person to feel this way. I understand why. There is a tremendous destructive pull when it comes to money. It can be used for drugs and prostitution. The mafia and cartels are fueled by money. People pay money to kill their enemies and commit terrorism. Money is at the root of child sex trafficking and a slew of the vile and monstrous things that human beings can do. However, many can also be used to fight these evils.

The truth is that there is nothing evil about money. Money is neither good nor evil; it's only currency. It's what you do with it that makes all the difference. You can commit terrible atrocities with money or do tremendous good. The choice is yours. If you seek to do good, you will eventually need to face down your money demons.

Hope

Hope is common in those who have yet to accumulate the amount of money they believe they need to be happy. In some ways, it's the opposite of the previous association because many people feel that their problems would end if they had just a little bit more money. *When I have $4 million in my retirement account, I can finally relax. When I get my dream house, life will be good. Once I get myself out of debt and pay off my credit cards, my life will improve. I'll be happy when I can finally* *afford to buy (fill in the blank with any of the typical things money can buy).* This is the idea that life will be better one day in the future when you finally get enough money.

Does it ever work out like that? No, because money doesn't fill the void. What happens is that the goal posts move, and we're left constantly wanting more. More money, more houses, more cars, more stuff. It never ends. Hoping is not a strategy. Wishing is not a strategy. Hoping and wishing without action are meaningless and delusional. But there is a way out of this type of thinking and all of the other traps laid by our money demons.

****

By no means are these the only four associations people have with money. Here is a list of some more common relationships:

- I'm always trying to get more money.
- Money equals status, prestige, and self-worth.
- There is a lack of fairness when it comes to money.
- I will run out of money.
- I will never have enough money.
- Money makes me stressed.

We can develop these relationships with money and many more from our family, friends, professional circles, society in general, the media, and our own experiences. Boil it all down, and most people feel money is hard to make, hard to invest, and hard to keep. The problem is that the money demons we develop from those associations bleed into every single thing we do.

We like to think we can compartmentalize our lives: we have our jobs, our marriage, our kids, our hobbies, and our money, and they don't ever intersect, but that's not true. Money is the thread that runs through every aspect of our lives. It literally touches everything, not just our investing and money decisions. It influences family vacations, charity donations, and the schools where we send our kids. It doesn't even end when we die because we have to figure out what we're going to do with all of this stuff that we accumulated throughout our lives when we're gone. That is why it's so important we have a healthy relationship with money. To think otherwise is pure fantasy.

How to Destroy Your Money Demons

You can't expect to create a new, healthy relationship with money until you destroy your negative beliefs and squash your money demons. That requires severing your attachment to the past by confronting the mistakes and breakdowns you've experienced with money and investing. This can be accomplished by following this seven-step process.

Step 1: List Your Current Money Beliefs

By this point, you're developing a better understanding of your relationship with money. So, to finally begin ridding yourself of those money demons, start by coming up with a comprehensive list of your current money beliefs. Write them down and get them all out on the page.

Step 2: Separate Your Beliefs into Positive Beliefs and Negative Beliefs

Remember that not all of your beliefs about money are bad, so I want you to separate the items you listed into positive and negative beliefs.

A positive belief is a pattern of thinking that gives you strength and the capacity to move forward. These beliefs create feelings of love, passion, energy, connection with others, and self-expression, so you're empowered to take action.

A negative belief is a pattern of thinking that reveals itself through ongoing worries, doubts, and complaints. These beliefs create feelings of pain, suffering, doubt, fear, inaction, resentment, anxiety, and jealousy, while making it difficult for you to take action because you're stuck in a spiral of negativity.

You can identify what type of belief you have by the emotion that it evokes. And those that create negative beliefs are your money demons.

Step 3: Determine the Source of Your Money Demons

These beliefs had to come from somewhere. How did your family speak about wealth or poverty when you were growing up? What beliefs did your parents hold? Which of those beliefs did you accept or reject? What lessons did you take from your experiences with owning cars, homes, buying insurance, using credit cards, and your interactions with banks? Determining how you developed a particular belief makes it easier to take away its power and weaken its hold over you.

Step 4: Determine the Price You Have Paid for Your Money Demons

Think about the negative feelings these money demons have created. They can destroy confidence and lead to anger and resentment. They can take any personal enjoyment out of activities that are supposed to be fun, as well as create stress, anxiety, and fear that negatively affect your health and well-being. If you think of money as a source of survival, it becomes difficult to enjoy that luxurious family vacation, no matter how nice.

Step 5: Consider the Emotional Gains These Beliefs Have Provided

This next step may seem counterintuitive, but it's crucial to sever the ties you have developed to your demons. Yes, there are hidden psychological "benefits," or secondary gains, you get from indulging your money demons, and these benefits can make it very difficult to relinquish these beliefs. For example, although these negative beliefs may harm your relationships with family and loved ones, they might create a sense of control, righteousness, and justification. They can be a way for people to avoid real communication. They can help you avoid responsibility for your money woes by blaming others and playing the victim. Acknowledge and confront these emotional gains because they can be addictive and a thorn in your side as you try to move forward.

Step 6: Create Guardian Beliefs for Each of Your Money Demons

A guardian belief is a positive belief that makes you feel strong and offers you the capacity to move forward. For example, if you see money as a source of survival and a struggle to accumulate enough to get ahead, use your guardian belief to put your situation into a different context. "I live in the greatest country on the face of the planet. I live in the greatest period in history to be alive."

If one of your money demons is paying taxes, and the idea of paying taxes creates fear and anxiety, flip that belief on its head: "Paying taxes is a patriotic privilege. I am lucky to have created so much wealth and income. I can double my income faster than the government can take it out of my pocket." This is one of my money demons that I've had to battle, so I understand the frustration you might have regarding taxes.

Step 7: Take Action

Beliefs are important but meaningless if you don't take action to support those beliefs. That's why the next step is to start taking positive actions to offset those money demons and reinforce your new guardian beliefs. That might mean setting up an emergency fund and planning a contribution schedule to ensure your family always has enough. It might mean calculating your taxes ahead of time and coming up with a financial plan to help offset the cost of taxes. If you're worried about being able to leave enough money for your children after you die, sit down with a professional to review your will and trust. Break the no-talk rule and instill proper money values in your kids so you can rest assured that when the time arises, they will be prepared. Don't run away from your money demons, or bury your head in the sand hoping they will magically disappear. Dismantle them by developing a plan and charging head-first toward a solution.

The Fear of Not Having Enough

Like the clown Pennywise, in the book *It* by Steven King, money demons can, and do, take on new and varied forms. The true underlying form of the evil entity stalking Derry Main is not revealed until the final chapters of King's grand opus. It is a shape-shifting monster from outside of our universe; a place known as the macro verse. It can take on any form, one that embodies its victims' most dreaded fears. It also feeds off the terror of its prey.

Although *It* can also kill and devour adults, *It* prefers to feed off children, and it savors the young and powerful fear their imagination ignites. If a child is most afraid of Frankenstein, then *It* takes that shape. If the werewolf terrifies a kid, then it assumes that appearance. Underneath it all is a malevolent creature from another dimension hell-bent on destruction. And although the monster terrorizes the city's inhabitants, it is perhaps billions of years old and seemingly impossible to kill. After an intense killing spree, it lies dormant for decades, lulling the town's citizens into a false sense of security, only to eventually reappear and begin its new reign of terror. It is the same with money demons.

Money demons have the baffling and confounding ability to morph into the thing that you fear the most. The thing you seldom admit to others and maybe not even to yourself. The equivalent to a mental "bump in the night," they can cause you to lose sleep and dread the future. They feed off your fears, real and imagined. They can lead to a mental state of being devoid of hope and vision. They can diminish self-esteem and freedom to be fulfilled in life. Like the monster *It*, they delight in the suffering of others. Although they are hard to "kill off," they can be defeated. But like monsters in a horror story, they don't always stay dead.

It may appear your demons have been vanquished, only later to discover that they have simply transmuted into another fear-driven nightmare—often more hostile than its previous embodiments. Money demons are powerful and patient. They most often show up as complaints people have about money or financial issues in their life.

Although they can take on myriad forms, I have found they have one underlying aspect in common. There is one belief that gives birth to all other forms. It's true nature springs from the filter called "I don't have enough." It is a sneaking premonition that often goes unacknowledged in the conscious mind but is always lurking hidden deep within. The internal self-talk sounds like this: *I don't have enough to feel safe, not enough to be happy, not enough to start my company, not enough to retire, not enough to create my dreams and not enough to do what I really want to do with my life.* It is just never enough.

Not enough, is the grand monster behind all demon beliefs, but unlike the monster in *It*, who hides and manically waits in the sewers, this life-sucking monster hides inside the subconscious mind, biding its time to take on new horrifying manifestations.

I have suffered the ravages of money demons in one form or another for most of my life, at least from the time I was aware of money. Today they can still plague me and cause potential conflict in my life. These destructive stories and what we make them mean can rob us of opportunities, creativity, powerful and fulfilling relationships.

Consider two I have done battle with over the years. These created massive frustration for me as I tried and failed early on in growing my business. They were particularly powerful in the mid-1990s, shortly after starting my company. I think you can easily see that anyone with these beliefs would find it impossible to scale and grow a company when you're constantly telling yourself, "It is impossible to find good people to work for me so I might as well do it myself" and "I am a poor manager, and it is hard to get people to do the right things at work." The short form was: good people are hard to find, hard to keep, and hard to lead.

Looking at the impact these morphing demons had on me and my company it became brutally clear that I would never be able to grow my organization to help more people and expand my reach. If I could not do battle with these inner lies, my dreams would be dead. Later, after I successfully killed these off, I discovered they changed into other fear-driven ghosts, haunting me with new self-destructive beliefs, more catastrophic than their predecessors.

To defeat them I had to first understand and dissect them. So, I needed to ask some good questions. What was I getting out of having these beliefs in place? There were two main payoffs, both disgusting to me as I identified them. First, they allowed me to play small and not have to risk hiring new people. I could be a martyr and complain that if I could only find the right people my company would take off. But make no mistake about it. I believed it was because they simply weren't out there, not because I was bad at finding them. After hiring several people who had poor job performance and subsequently needing to let them go, I created the screen that "I could not manage people." That screen allowed me to avoid tough conversations with employees when they needed coaching to elevate their performance. The bonus payoff was that I got to "be right" and blame everyone but myself for the fact that my company was not growing at that time. Juicy stuff, right?

After analyzing the cost and payoff of my money demons at work it was simply unacceptable to keep them in place. I was not willing to sacrifice my dreams on the altar of fear, self-pity, and self-righteousness. I had to first destroy the old meaning I had given my past experiences and develop alternative screens, or guardian beliefs. It simply could not be the case that there were no great people out there. After all, some companies had thousands of employees. They couldn't all be incompetent.

It had to be the case that there were good people and even awesome people out in the world who were dependable, talented, and committed, I just had to find them. And, of course, I could manage and train people. It just meant being strong enough to have real and authentic conversations with them and caring enough to help them identify and reach their dreams.

My new beliefs became, "There is an abundance of amazing people in the world who desire to live an extraordinary life and become leaders and are out to make a difference in the world. After I recruit them, my job is to keep them motivated, committed, and communicating with each other authentically and build them into a world-class high-performance team." This was my new guardian belief, one I nurtured and focused on every day.

In time, the new screen was firmly in place and enabled me to expand my business exponentially. Today I have more than 70 employees pulling for our mission and values as a company. What was once one of my greatest drudgeries, managing employees, is one of my greatest joys as an entrepreneur. It is a great honor to lead my team and work with them shoulder to shoulder. I am constantly in awe of their creativity, dedication, and innovation. Instead of me lifting them up, it is quite often the opposite: they constantly inspire me to evolve and expand both myself personally and the company's capabilities.

The camaraderie and teamwork that was born from these new beliefs left me feeling moved, touched, and inspired. I dismantled and destroyed those twin money demons decades ago. But like the monster *It*, they were only partially dead, and when they reawakened, they did so with a vengeance. The new demon was one so powerful it had the power to destroy any chance I had of ever being in a loving marriage. Yes, unchecked they have that kind of power.

After my divorce, I am loath to admit I was overwhelmed by a pervasive fear that I would never be in a loving relationship and was plagued with perpetual loneliness. The apprehension of this possibility was palpable. This underlying agitation fed into my childhood terror that I was not attractive and therefore undesirable. Although I can see the inner workings of my subconscious today, they remained largely hidden from me at the time.

Based on my divorce and chaos in relationships, the new and devastating money demon from the past were reincarnated into new shapes, which sounded in my head something like this, "Being in a trusting, loving relationship is impossible because no one will want you—just your money. Therefore, you will never be able to trust any women you are in a relationship with." Money demons don't just affect financial dealings; they almost always have a relationship component as well.

They obstruct our ability to share openly, honestly, and freely, due to the fears of rejection, jealousy, greed, and revenge they often stoke. Money demons have been the sad cause of many marriages being destroyed. Jealousy has obliterated rock bands and businesses alike. But it should be noted that the money itself does not cause this pain and suffering. No! It is the story we tell ourselves about money and what we make it mean.

Ironically, when my wife and I met she also had a money demon about relationships. Hers was, "Men who have money are arrogant, egotistical jerks." Ouch! She was reluctant to give me her phone number when I asked her out to coffee at Starbucks, and I was reluctant to ask her for it. But because I had done battle with these types of demons before, I knew there was more than a possibility that my story, "all women could not be trusted," was wrong. And I had been visualizing the type of woman I really wanted to fall in love with for some time (more on how to do that in Chapter 9). So, we both chose to fight off our first instinct about dating and openly talked about our fears.

I have discovered that money demons linger in the dark and (like vampires) hate the light of day. The best way to slay a money demon is to share it openly and honestly with the people who mean the most to you. In no small way, the demons are evil because they rob us of the love and life we might have lived and they thrive in obscurity, hidden deep underground, inside the mysterious universe governed by the no-talk rule.

To the extent you can identify these shape-shifting tormentors, break the no-talk rule, and share them graciously and kindly with others while developing guardian beliefs you will expand your capacity to live an extraordinary life. But fair warning, I've learned that when I am in their grip, the fear I need to express hides and in its place anger often emerges. This type of anger and resentment destroys closeness and affinity in relationships. It tends to isolate me and make me feel that I am alone.

As you expand your leadership, create your team, play a bigger game, and even accumulate greater financial success be warned that your money demons will not shrink away and softly die in the corner of the room. Oh no, they tend to be the opposite. Your demons grow right along with your success. You must be constantly vigilant and ready to do battle. Your American Dream demands it.

Money demons can be self-fulfilling prophesies. The stronger I believe something to be true, the firmer its grip on me—and I act accordingly to prove my inner voice right. Acts that are inconsistent with these beliefs are unthinkable. I create more of the thing that I fear the most, and at that point in my life, it was a deep-seated fear of abandonment.

My money demon had transmuted into a relationship demon that nearly cost me the love of my life and the amazing family I cherish today. Perhaps it was this same type of demon that plagued my grandfather, George. It seemed there was a series of thoughts in his mind that kept him a prisoner in his own head.

Like a chain reaction, one that can affect us all, something happens, we give meaning to it, we start to believe that it is "the truth" about ourselves, or worse, "the truth" about the world. And that "truth" dictates how we live our lives. The thought of that is far more terrifying than any book I have ever read or movie I have ever seen.

Doing battle with your money demons is the first step to improving your relationship with money for you and your family. But it's important to be patient with yourself and the process. You've fed and nurtured these demons over a lifetime. You can't expect to get rid of them overnight, but you can get rid of them. One of the best ways to accelerate this process is to change how you think about money and see it for what it is, because it's not what you think.

Money Is a Phenomenon of Language

Many people have the idea that they must fight to obtain and accumulate money and protect it from those who want to take it away. As mentioned, people believe money is hard to make, hard to invest, and hard to keep. That is all based on the idea that money is a scarce physical resource. Physical resources are limited. But money is not physical, nor is it limited. It's a phenomenon of language.

Money is part of our social structure for managing and creating the success of our species. Your dog doesn't know what money is. Your dog doesn't have an IRA or a checking account. Your dog doesn't care if you wear a Brioni suit or a T-shirt and jeans when you walk it. All my dogs want is a treat. And to lie on top of me when I watch television. They don't care about money. A $100 bill is not a $100 bill to your dog. That's because money only exists in a conversation.

You might argue. "You sound crazy, Mark. Of course, money is real. I can physically hold it. What are you talking about?" Yes, you can hold physical money, but why is that the case? We agree that certain pieces of paper have a certain value, and we can trade those pieces of paper for products. That agreement is created through language. Give that $100 bill in your pocket to a member of a Brazilian tribe, and it would be worthless to them. Why? Because we haven't communicated any agreement with that tribe. Those tribes might measure wealth by how much cattle you own, and if you don't own any cattle, they might consider you poor. Money is not universal. It's a conversation between two human beings. So, if you change your language around money, you can change your relationship with money.

Try thinking of this another way. The vast majority of your money is not in your physical possession. You might not even have $100 in cash on you right now. Where is all that money? It's electrons flying around somewhere in the cloud. It's numbers on a screen. Chances are that you will probably never touch 99.9% of the money you earn in your lifetime. It will never even be exchanged into paper form. Yet, we've agreed that those electrons are worth something specific. It's strange when you think about it like that.

Look back over the list of how different people talk about money. Everyone said something different when I asked them to tell me about their relationship with money. Talk to 10 people about what money means to them, and you'll get 10 different answers. So, who's right? They're all right because that's what money means to them, so that's how it shows up in their lives. That's what most people don't realize. My grandfather thought money was evil, so only evil people could have a lot of money, and guess what? He never earned a significant amount of money. Because if he did, in his mind that would make him an evil person. He literally died to keep that belief about money in place. Meanwhile, my father believed that earning money was a sign that you were helping other people and making a positive impact on the world.

You can master any new subject by mastering its conversational domain. Money is a phenomenon of language. Who you are as an investor is also created through language. That means transforming the way you see investing comes down to mastering new forms of communication and language. However, you must choose your words carefully because they can influence your thinking. This occurs subconsciously so most of us never know it's happening, but our thoughts shape our words. Our words create our world because they influence our actions and determine our results. By taking back control of the language you use about the subject of money, you can create new possibilities and get new results. Human beings create their reality through language, so to master anything, you must master the language of it.

Money is what you say it is, so what you say about it becomes your default position. But because money exists only in conversation, that means you can change the conversation. Why not make it something empowering? Make it something that can help you create freedom, fulfillment, and love. Once you understand that is possible, you can transform your relationship with money forever. My father used to half-jokingly tell me, "If you want to see how many people you're helping, look at the top of your income tax return." What he meant by this is that the more people you help the more money you will inevitably make. For him, and now for me, it is the opposite of evil; it represents the value you create for others. This screen has kept me perpetually seeking new ways to create a higher quality of life for other people. This is a powerful screen for living an extraordinary life.

CHAPTER

Create Your True Purpose

"Money has never made man happy, nor will it. There is nothing in its nature to produce happiness. The more of it one has, the more one wants."

—*Benjamin Franklin*

"What is your true purpose for money?"

Early in my career, I always operated under the assumption that if you had more money, you had more security, more peace of mind, and lived a better life. I believed that more money would make you happier. And if you had less money, the opposite was true—you would be more inclined to be fearful, angry, upset, and afraid. So, I saw my job as helping people accumulate wealth. By doing so, I believed I contributed to their happiness. Two of my clients blew that theory completely out of the water.

At 27 years old, one of my richest clients, let's call her Mrs. Smith, had millions, which made her a pretty good client for a financial planner like me to have so early in my career. Despite being my wealthiest client, she was not secure or happy. Far from it. She was my most miserable client. Having lived through the Great Depression, she was terrified that the market would crash, and she'd lose all her money. That was her money demon. Money was associated with fear, and it's how she lived her life. She wanted to hoard as much money as possible to avoid running out. And that pursuit was completely self-serving.

Another client, let's call her Mrs. Jones, had significantly less money than Mrs. Smith. However, Mrs. Jones had something that Mrs. Smith did not. Mrs. Jones was happy—like amazingly happy. She spent her time with friends and family. She went on nice vacations and donated money to charity. If the market went down, she'd find a positive spin. "This is great. We get to buy more when it's down. Right, Mark?" She never left my office without a smile.

I had a difficult time wrapping my head around this realization. I was experiencing cognitive dissonance, and that led to an existential crisis of sorts. If I spent the next 30, 40, or 50 years as a financial advisor counseling people on how to make money only for them to become unhappy—well, I couldn't live with that. What good would I be doing if I helped people make money and didn't improve the quality of their lives? Worse, what if I helped them make money and they became miserable? That would be a complete waste of my life. I needed to figure out why some people were happy, and others were not, despite the size of their portfolio.

In *Think and Grow Rich*, Napoleon Hill wrote about finding a vision and creating a powerful purpose for your life. What reinforced this concept for me was a book called *Man's Search for Meaning* by Holocaust survivor Viktor Frankl, which I was reading at the time I wrestled with this dilemma. I realized that so much of what I experienced with these two clients traced back to purpose. Mrs. Smith apparently had no purpose and Mrs. Jones did. That's when it finally dawned on me: money doesn't make people happy. Living with purpose does.

A famous 1978 study[1] compared the happiness of lottery winners to that of paralyzed accident victims. Most people would assume that the lottery winners would see a significant increase in happiness and the victims of accidents would become significantly unhappy, but that was not the case. The level of happiness hardly changed in either group studied. In other words, when the shock of their new circumstances wears off (either positive or negative), people tend to return to their default state. Just because you experience a financial windfall, doesn't mean you become any more alive or fulfilled. It doesn't mean you become closer to people. It doesn't mean that you become any more passionate, joyful, or loving. Your life doesn't become extraordinary through money alone.

Not only does money not make people any happier, it also can have the opposite effect. It can contribute to making them unhappy because they start to worry more and fight with their loved ones. Money can create jealousy, anger, and resentment. Others, like Mrs. Smith, can become incredibly fearful that they will lose their money, and spend so much of their time and energy trying to hold onto or hoard their money. They strive to make money for the sake of making money. Meanwhile, they remain trapped in the world of "not enough" and constantly chase the idea of more. It's a trap that I suspect that we've all experienced at one time.

The Destructive Cycle of Wealth

Yes, it's a cliché that money can't make you happy. Songs have been written about it. But why can't money make you happy? Shouldn't it make you a little happier if you suddenly stumbled into, say, $100 million? The answer is no. It sounds counterintuitive, but I firmly believe that to be true.

If money isn't making us any happier, and most of us have much more than we need, why do we work so hard to make more of it? Why do we spend so much of our lives thinking about money? Without even realizing it, we are stuck in the destructive cycle of wealth. Learning about this cycle can help you better understand some of your decisions about money. It has five phases, beginning with our most basic human needs.

Phase 1: Survivorship Stress

As human beings, we all have an instinctual drive to survive. We instinctively protect our family and strive to help them survive as well. That's a big responsibility, and in the modern world that responsibility requires money.

What do you need as a human being? Strip life down to its essentials, and you need food, shelter, and clothing. That's all normal and rational, so we go into the world to obtain items that fulfill those basic survival needs. So far, so good. It can be stressful, but nothing is wrong or unnatural about this phase when approached from a realistic perspective.

Phase 2: Human Wants

Once we have met those basic survival needs and discover that we have some money left over, we tend to use it, not for what we need, but for what we want. It might be a new purse or set of golf clubs. Maybe we want that new iPhone. You don't absolutely "need" this stuff, but you "want" it. I consider a want anything more than utility, or the bare necessities required to live life. If I need a pair of shoes, I can go down to Payless and get a $50 pair of shoes. I don't need a $3,000 pair of shoes from Neiman Marcus. I can find a basic car, but I don't need a Ferrari. I need clothes, but I don't need nearly as many clothes as I currently own. Those are all human wants. We've all wanted things that go beyond basic utility, so you know exactly how that materialistic urge feels.

Phase 3: Obtainment

We don't always act on our wants and desires, but when we do, we obtain those possessions. I will assume that anyone reading this book isn't living off the earth with only a few meager possessions. We've all spent our hard-earned money to buy things we want. Pretty soon, the things we obtain get a little bit bigger. It goes from purchasing that new computer to getting a bigger house and a nicer car. Some of us have so much stuff that it doesn't even fit in our house, so we have to pay for public storage space to put it all. And we still want more!

Phase 4: Relief and Exuberance

Getting what we want feels good. I had dreamed of living in Florida and having a boat for most of my life. I worked hard and got to the place financially where I could finally do it. I purchased a 37-foot boat. It had three Yamaha motors and was great for fishing. I loved that boat, and it gave me this amazing sense of relief and exuberance. I had my dream boat and could take it up and down the Intercoastal Waterway. It was exciting when I bought it, at least for a little while.

Phase 5: Comparative Analysis

That brand-new thing always feels good initially, but it doesn't last. That's the problem with material possessions and most things we can purchase with money: that good feeling these items create is only temporary.

One day, I was out on my boat, and some guy sailed by with a 150-foot yacht. That's when I noticed that the tender he used to get from the dock to his boat was bigger than my 37-footer. *I need to go get a new boat!* Up to that point, I really did love that boat. It brought our family close together, and we experienced such great times together, but that shift in my mind occurred almost instantly when I saw that bigger, better boat. It's foolish, but it's a human instinct.

We've all been in similar situations where we work hard to get what we want, and we feel great about that purchase, only for that excitement to fizzle and fade. You can purchase your dream house. Every room is perfect, and you feel that you'll never have to buy another house ever again because this house fulfills your dreams. And then one day, you look around and say, "These kitchen cabinets are kind of crappy. Look at this tile! That's so 2000s. I don't even like this color. What was I thinking? I can't live like this. Let's rip it out and put in a new kitchen."

There's always someone with a bigger boat, and a better house with a nicer kitchen. There will always be a nicer purse and a better pair of golf clubs. Think about every time a new iPhone comes out. People run out to get it because this one is encased in titanium and can take pictures in the dark. No matter how much you fall in love with those material possessions when you first get them, it's only a matter of time before you want that new, bigger, and better thing. That is the destructive cycle of wealth. And you probably don't have to think very hard to recall when you found yourself caught up in it.

This cycle is so destructive because none of these material possessions will ever make us happy. I call them cookies and toys. Cookies are the luxuries you ingest, and toys are the material objects that make us temporarily feel better. And I love cookies and toys, but I've also learned that even though all of these purchases can feel great at the moment, they don't provide true, lasting happiness. And they certainly don't provide peace of mind. This is true of every feel-good stimulant, whether you buy it, drive it, wear it, or ingest it. Not only do these things not provide happiness, but if consumed in excess, they can have negative effects on our well-being. Somewhere down the line we all pay for our satisfaction.

That is why I can unequivocally say there is zero correlation between having a lot of money and a high degree of happiness. I know this to be a fact because there were times when I was dead broke and happy, and there were times when I had a lot of money and was miserable. True happiness comes from values realized in your life. It comes from your sense of purpose. Money is a tool. When used properly, it helps you to accomplish the things that are the most meaningful and important in your life. The sooner you grasp that, the easier it will be to connect with your purpose and what truly matters.

What Is Purpose?

I was sound asleep late one night when the house alarm woke me up. Still groggy, my first thought was that someone was trying to break into the house. That most likely meant someone was trying to get in through the garage on our property. Without thinking, I jumped out of bed and ran to the garage. I knew any intruder might have a knife or a gun, so if I were to get into a confrontation, I knew that I could die. In a flash, I thought *I'm good with that.* My wife and kids were in the house. Even if I couldn't stop the intruder, I could at least slow them down, so my family could get safely away. *If I go down, I go down, but I'm not letting that person get into my* house. It turned out to be a false alarm, but I initially felt that way because I had a purpose—to protect my family. And that purpose was greater than life itself.

When you have purpose, you are driven to do anything that aligns with that purpose. Isn't it ironic that the thing you take a stand for and are willing to die for is actually what makes you feel alive and makes life worth living?

So, what exactly is purpose? It's *not* the pursuit of pleasure and self-indulgence. Although those endeavors may make you happy in the moment, they are temporary, short-term pleasures. Purpose is the reason for which something is done or created. It's directly connected to values and character. It's part of your very DNA. It's what aligns everything in your life. Purpose can transform your state of being and your reality. It helps to fashion your relationships and gives you direction and meaning when moving through life.

Many people resist finding their purpose—they just want to make as much money as possible. If you're an accountant, engineer, financial planner, or a numbers person who is highly analytical, you might just want to see a spreadsheet. Forget purpose. The formulas are your purpose. Others think purpose is touchy-feely and will make them soft. Some even express pride in the fact that they don't show emotions or feelings. That prevents them from getting in touch with what truly matters in their lives.

Purpose can be elusive for some because they struggle to find meaning when so much suffering exists in the world. When I catch myself thinking like that, I refer back to Viktor Frankl. While a prisoner in a Nazi concentration camp, Frankl lost his entire family. He witnessed an unimaginable amount of suffering, torture, and death. Amidst all that hopelessness, Frankl observed two types of prisoners:

those with purpose and those without purpose. Those without purpose died quickly. Although purpose didn't alleviate anyone's suffering, it enabled prisoners to find meaning in their suffering. If you can't find meaning in the midst of suffering, you'll never find meaning, because all human beings suffer. If Frankl could find meaning in that scenario and live a fulfilling life, then he was definitely onto something with regard to purpose. Yes, there is tragedy and suffering in the world, but that's when purpose is needed more than ever to serve as a guide. That will help you succeed and keep on going when outside circumstances destroy others, because with purpose, you can make something possible that was never possible before.

Without the direction that comes from purpose, it's easy to become lackadaisical, nonchalant, complacent, and even apathetic. It leaves you putting off doing what you need to do until the future, because you assume life will get better. But that is rarely the case because, when you are in that state, you lack the motivation to change your current predicament. That leaves you trapped in an ordinary life. Unfortunately, that's reality for too many people.

Purpose is about taking responsibility for your life and your future. In other words, it is the opposite of victimhood, which is where my grandfather was stuck for his entire life. Those who see the world through the screen of victimhood feel that others are responsible for their circumstances, which can't help but lead to anger, resentment, and jealousy. Victimhood never leads to positive results. You absolutely must be responsible for your own life, and purpose is your entire reason for living.

Your purpose in life is also connected to your purpose for money. That means it's powerful to invest your money in alignment with your purpose. If you are trying to pick the best stocks, get in and out of the market at the right time to optimize returns, and find some five-star manager to give you all the hints and tricks, you are speculating and gambling with your money. That is putting your family's future and your American Dream at risk. It's in direct opposition to your purpose. Once you can make that connection, you have taken the first step on your way to creating freedom, fulfillment, and love in your life.

When you don't have a purpose for your life, you don't have a purpose for your money. Or worse, you have a false purpose—and you simply want more money, and a bigger, better portfolio. We think that more money will make us feel better, and unfortunately that's the default state for money people. When in that state, money is destructive, and it becomes a corrosive thread that runs through every aspect of your life. It can make you unhappy and even miserable as you struggle to escape the destructive cycle of wealth. Without purpose, no amount of money will ever be enough, and you remain doomed to repeat the same mistakes of the past. However, if you have a purpose, any amount of money will do because you're in pursuit of something much more meaningful.

The reality is that as an investor, you will take financial hits at times. The market goes down as well as up. Outside events will affect your self-esteem and willpower. Without purpose, the default screen, or your old relationship with money, will kick in, and the results can be extremely disempowering. Purpose enables you to take that power back. It helps you escape the destructive cycle of wealth so you can invest and live on your terms. That's what brings your American Dream to life.

The Power of Purpose

Everyone who has achieved greatness or accomplished something significant, or perhaps even changed the world, has done so because of purpose. Gandhi's purpose was to achieve freedom for his people through nonviolence. He would not have been successful and changed the world if he hadn't had a clearly defined purpose. Think about the historical figures you admire. People like John F. Kennedy, Abraham Lincoln, the Wright Brothers, or any leader or inventor, builder or artist who did great things, and you will see that they were all driven by deep purpose.

I can promise you that there is nothing soft about purpose. The opposite is true. Nothing takes more courage than having purpose. People have died for their purpose. Purpose has liberated the enslaved and thwarted despots. Purpose is the ultimate context for viewing the world and how you fit in it. No purpose, no change. No purpose, no power.

"Life is either a daring adventure or nothing," Helen Keller once said. Her life and words influenced millions. "I have a dream that my four children will one day live in a nation where they will not be judged by the color of their skin but by the content of their character," Martin Luther King said in his most famous speech. He changed the world with his purpose, and he actually died for it. That's the ultimate sacrifice. Ronald Reagan, in his famous Berlin Wall speech in 1987, said, "If you seek peace, if you seek prosperity, if you seek liberalization, tear down this wall." His purpose was to spread the power of freedom throughout the world.

Purpose is not a soft science. The subject has been researched, and the results are shocking.[2] Having a definitive purpose in life has been linked to increased health, prosperity, and well-being. Purpose has been shown to reduce pain, fear, depression, stress, and loneliness while increasing optimism. It helps us repair our DNA, improves the quality of our sleep, increases good cholesterol, and promotes cancer-killing cells, while reducing inflammation and reducing hospital stays. More significantly, it has been shown to increase lifespan. One study looked at individuals over 50 with coronary heart disease and found that those with higher sense of purpose had a reduced risk of heart attack after a two-year follow-up.[3] Another study showed a direct correlation between purpose in life and physical health and fitness in older adults.[4]

Purpose has been shown to increase motivation and optimism and make the task in front of us seem less daunting. In a fascinating 2016 study, participants were tasked with walking up steep hills. Researchers discovered that those with purpose underestimated the steepness and effort required.[5] In other words, operating from an elevated state of purpose better enables us to view daunting tasks as more manageable.

I placed this chapter in Part I of the book because you must find your purpose before I try to talk to you about making any investing decisions or life changes that may conflict with your current beliefs. As you will learn (or you might have already experienced), behavioral changes are difficult. Whether trying to create more positive habits about our health, fitness, diet, or investing, it's difficult to challenge existing beliefs. It's all too easy to revert to our familiar and comfortable default habits. When those beliefs are challenged, it's natural to experience cognitive dissonance and become defensive. However, people are more willing to change their behavior when operating from a place of purpose. When purpose is bigger than simply making more money, it reduces ego and makes change more palatable. It can pull you away from the default state. Why? People become less concerned about defending their beliefs and more open to doing something that could help them achieve their purpose. So, when I show you Nobel Prize–winning investing research in Part III, you will understand how speculating and gambling with your money is not aligned with your purpose and will be more open to change.

If you don't have a true purpose, your default purpose becomes survival and more money. That can lead to an ordinary life driven by scarcity and fear. That's why most people don't wake up excited to crush the day. They don't tell themselves they intend to be the best parent in the world and make their spouse feel like the most important person on the planet. They don't think about creating value for others. And yet you must create value for other people before you ever expect anything in return. That is a lesson that has guided me since college and through every step of my career. Because I have purpose, I wake up every single day thinking how I can be of service to others. I know that if I can help enough people, I won't have to worry about money. That will take care of itself. The money will come. And as my father used to say, "The money is the measuring stick to determine how you are being of service to others." That's the type of mindset shift that occurs when you have purpose.

How to Discover Your Purpose

My purpose is to empower families to discover their true purpose for money. I achieve that by transforming their investing experience, leaving them with freedom, fulfillment, and love. I love to watch people become stronger and more confident, especially when helping them take more responsibility and seize opportunity. That is at the heart of what I hope to accomplish. It's not a slogan or a catchphrase. It didn't just appear to me in a dream. I spent a lot of time developing this purpose—it began back when I was 14. You should do the same with your purpose. So, roll up your sleeves and let's flesh this out. I don't expect you to discover it right away. Like anything worthwhile, it will take time and effort.

I can hear all the questions you want to ask already: *How do I create purpose for money? Where do I find it? Where do I even look?* Well, purpose is not something that's already in in the outside world. Your purpose doesn't exist yet, which means you must create it.

Start with a brainstorming session. Consider everything that's important in your life. What do you value? Take a pen and paper and write out as many possible purposes as you can think of. What could you do with money that would make a difference for yourself, your family, and for others? Purpose is more powerful if it doesn't try to fix a problem. Don't focus on a negative or on what you're trying to avoid; that will only create more fear and panic. You want your purpose to be positive. Focus on what you're moving toward and trying to achieve for others.

Don't analyze your answers yet. Just get them down on the page. If you're struggling to come up with ideas, just start writing. It doesn't matter if an answer is big, small, silly, or feels wrong. There is no one right answer for this. It's completely up to you. Here are some examples to get you started:

- To inspire in my family and community the power to create.
- To empower people to thrive and grow.
- To leave a family legacy of generosity.
- To create faith in family and the world.
- To create love and connection in my family.
- To create love and service to my community.
- To support causes that make a difference in the world.
- To have my life be useful to humanity.

- To live a loving, adventurous life.
- To help foster kids who have aged out of the system create worthwhile lives.
- To experience wonder.

During one advisor workshop, somebody stood up and said that their true purpose for money was power and control. I asked him what he wanted to do with power and control, thinking he would use it to help other people. Much to my surprise, that's not what he said. He actually wanted power and control for the sake of power and control. I had to ask him to leave because my true purpose was to create freedom, fulfilment, and love, which is the complete opposite of what he wanted. I had absolutely nothing for him. But at least he admitted it. Most people who are driven by greed lie about it and claim to have altruistic intentions when they are just trying to make as much money as they can as fast as they can.

Once you have your list, it's time to analyze what you wrote. Go through each one and ask yourself, *Does this inspire me? If I lived life and used my money to achieve this purpose, would it lift me up? Would it create freedom, fulfillment, and love in my life, for my family, and for the people I care about?*

If you can answer yes to these questions when analyzing one of these statements on your list, circle it. This will help you identify the statements that have the most powerful and positive impact. Share these statements with those who live inside your American Dream. See how it feels to say it out loud and tell someone else. Try it on like you would a new pair of shoes.

Some people find a way to incorporate their purpose into their work. I have a client who is an optometrist, and his purpose is to change people's lives through vision. If your purpose and career are aligned, one thing you want to consider, especially if you're getting older, is what you will do when you retire. I sometimes tell people that one of the worst things they can do is retire. There is some truth to that because some of the happiest people I know are those who continue to work while in their 70s and 80s. They may slow down, but they're still growing and being of service to others. The minute you stop pursuing your purpose and start sitting at home watching TV all day, everything goes downhill. Consider that when crafting your purpose.

Others make the mistake of being too specific. When that happens, your purpose runs the risk of becoming too similar to a goal that can be achieved, leaving you looking for a new purpose. The broader your purpose, the more powerful and impactful it can become. Family is important, so that's where you want to begin, but also consider friends, career, and community. The further out you get, the larger your footprint and your potential impact. If you want to create meaning in your life and live a powerful existence, consider pursuing a purpose you would die for.

If you have a partner, work to align your purpose with your partner's. A client explained to me how she was always out of step with her husband when it came to money because she made more than him. That affected how she spent her money. She felt guilty and would curtail spending money on trips because of how she imagined her husband felt, when it turned out he didn't have the same feelings at all. This couple's first step to improving their situation was to eliminate the no-talk rule. The next was to align their purposes. If you're already in a loving relationship, your purposes are most likely already in alignment.

My wife and I have a similar purpose, and together we make an excellent team. She's strong, brave, and passionate about what we're working toward. She has my back, and I have hers. She sees things that I miss, a result of us being in complete alignment with the future we hope to create for ourselves and our family. As a result, both our marriage and our business have become more resilient. I can promise you that life gets really exciting when you can start living a powerful, full, and fun life with the person who is most important to you. When this occurs, it doesn't matter who makes more because money is no longer what's most important.

Gravitate toward people who have a similar purpose, and avoid those whose purpose is counterproductive as best you can. For the most part, you don't need to make an effort to do this because it will occur naturally. Occasionally, you will have to push people away or draw a boundary, but like-minded people are organically drawn to each other. Keep in mind that you can always refine or add to your purpose. You can look at what works and what doesn't, and proactively remove the imperfections. I've watched clients refine their purpose over time when they realized what they were working toward could have a much bigger impact if they simply broadened their purpose.

When You Struggle to Find Purpose

Some people's purpose is obvious, and they don't need to think much about it. If that's you, congratulations! For others, it's elusive. What makes finding purpose so difficult is that you can't force it. You must fully believe in it with every fiber of your being. It also can't be a manipulative tool you use to make more money or have a better life. It's not a strategy for how to win or get more. It must be an expression of what you believe is important in the world and what you are willing to take a stand for. If there is one way to expedite your search for purpose, it's this: get interested and committed to helping those living inside your American Dream achieve their purpose.

Almost all people have a purpose that is bigger than themselves. Purpose typically means serving others, so the more you focus on helping others, the less you focus on trying to get what you want. That gets you out of a selfish, scarcity mindset. Focusing on the self never works, just like trying to increase happiness will not create more happiness. Once you align your behaviors, relationships, and what's most important in your life with helping others, you will have an energetic, abundant, and rewarding life, no matter how much money you have in the bank.

I'm happiest when I'm teaching in front of a class. That's when I dip into what psychologist and author Mihaly Csikszentmihalyi calls the *state of flow,* and everything else in my world disappears. Time morphs, and I get to experience a feeling of expression and creativity. The negative thoughts are greatly reduced if not shut out completely when in that zone. Living an extraordinary life is all about creating those moments. That's when I'm at my most productive. More important, that's when I experience true freedom, fulfillment, and love. That's when I'm living my purpose, and it occurs by being of service to other people. That's why I say that purpose can't exist in selfish isolation.

Think of it this way: if your purpose is rooted in generosity, caring, and having an impact on others, and you have a friend in the hospital, you can be there for that friend in ways that no amount of money could ever duplicate. Yes, money can help, but it's not what's most important. Money is fuel for your purpose, but it's not a prerequisite. But just because your purpose isn't rooted in selfish pursuits or getting more money, doesn't mean you can't have nice things. However, you must understand that those things can't replace purpose.

It's ironic, but to create your true purpose for life (and money), you want to create a purpose that is greater than money. Do that, and money will loosen its grip on your life. You may not believe it at first, but when you are driven by a purpose greater than money, you frequently make more money. In my experience, that is often true. But there is a caveat. You can't hack the system. You can't just come up with a purpose or say that you want to improve people's lives. You absolutely must believe it to your core. If you had the choice between your purpose or having a boatload of money, and you chose purpose, then you know you've arrived. You can't take the money and hope to find your purpose, because you will still be trapped in the destructive cycle of wealth.

Don't worry; you will know when you finally discover your purpose, because it will hit you like a bolt out of the blue. If your purpose doesn't fill you with passion and power, chances are high that you haven't discovered it yet, so keep looking. But choose carefully. We are who we choose to be, so choose your purpose with purpose. The two most important days in your life are the day you were born and the day you find out why. Once you declare your true purpose, the real work begins, because that is when you incorporate your purpose into everything you do.

Taking Action

When I created the American Dream Experience, my goal was to help families achieve freedom, fulfillment, and love. I worked with my team and my wife Melissa to create the class. At the time, we were also enrolled in some personal development training. During the workshop, we discovered that we both still had issues with our former spouses. The leader of that workshop brought this problem to our attention. At first, I didn't know what she meant.

She told us, "You're talking about helping families achieve freedom, fulfillment, and love, and you both still have an ugly relationship with your exes. You both have kids with your exes, so they are still a part of your family. You're not working to get what you say you're taking a stand for."

That hit me like a slap in the face. "So, I have to stop criticizing my ex-wife?" I asked.

"That would be a good start, but you both need to mend those relationships."

She was right. I had anger and animosity that went back years. I knew it wouldn't be easy, but I had to make things right. So, I called my ex-wife, and we talked. I asked for her side of the story, and I listened. I had no outside agenda; I just wanted to better understand where she was coming from. I told her, "I want to have a better relationship for the children."

I then phoned Melissa's ex and told him, "I've been awkward around you because I'm jealous. I think I'm above it, but I haven't been. You two have a daughter together who is now my daughter, too, so I want to be a good co-parent with you."

Melissa also called my ex and her ex. We each took action to improve our relationships with our exes, and today, we all get along exceptionally well. Our kids have since come to us and told us they are glad everyone gets along because it weighed heavy on their hearts when there was friction. I had no idea that was happening, so it's great to know that we have gotten past it all. And when we did, it created a closeness in our family that didn't exist before.

Now, when I say I believe in freedom, fulfillment, and love for families, I practice what I preach.

Everything we've discussed so far will require action and hard work if you want to experience change. It begins with the questions you ask and striving to learn what you don't know you don't know. It continues by transforming your relationship with money. That means eliminating the no-talk rule and doing battle with your money demons. Remember that money is not scarce or a means of survival, but a phenomenon of language and a tool that can be used to create an extraordinary life. You then must work to eliminate your existing destructive screens and replace them with empowering screens, especially the screen of the American Dream. That will help lay the groundwork, so can start pursuing your purpose.

Purpose isn't wish fulfillment. You don't sit back and hope that your new purpose for money will guide you. You must be proactive. If language is what first plants the seeds for this new purpose, it's through action that you bring it to life. If you don't take action, you won't get results. Without action, purpose is a platitude, "a remark or statement, especially one with a moral content, that has been used too often to be interesting or thoughtful." In other words, it's empty. It lacks meaning or sincerity, so even though it may sound good, it doesn't have any impact.

I've learned that some people push back against purpose and put off having to find it because they intuitively know that it will require a commitment. And they are right. Purpose requires action, and some people subconsciously fear that. They might like the *idea* of changing their lives, but don't want to do the work or put themselves out there because they are content or have grown comfortable with their victim-driven existence. That's an obstacle that many people struggle to overcome, but those who can make it to the other side experience tremendous rewards.

Once you start living through your true purpose for money, your life experience often shifts from scarcity to the possibility of abundance. You will make money mean something, so you might as well make it something powerful and amazing. You attach value to it. When money becomes something empowering, that can inspire and lift you up.

Start by looking to see where your purpose is missing from your life. Look for actions and behaviors that are inconsistent with your true purpose. Ask yourself: *Where do I feel constrained? How do I feel fearful? What have I not acted on that I previously committed to?* Look back at broken dreams and broken relationships. You should do this without any judgment. You just created this purpose, so you should expect to find gaps. It would be unusual (if not impossible) to discover your current life lined up perfectly with this new purpose.

It doesn't matter how far your current life might be from your purpose. What matters is that you are willing to take the actions required for it to show up from now on, going forward. Are you willing to take that step? If so, it will open you up to a whole new world of possibilities and opportunities.

Look honestly at your daily behavior. How often do you take the necessary actions to live according to your purpose? Let's expand the question from the beginning of the chapter and look at it with a fresh set of eyes.

What three actions can you take this week that would be consistent with your true purpose for money?

Consider what actions or relationships would give you power over money and help bridge the gap between where you currently are and your true purpose. Think about what you could do for others to help them realize their purpose and the American Dream. It might be a phone call, a meeting, or a simple act of kindness.

There are no right or wrong answers. The point is to tap into the power of your subconscious mind, because it can lead you to discover things lingering beneath the surface that you haven't quite realized yet. If you want to take this commitment to the next level, share your purpose, and three intended actions consistent with that purpose, with those you love. When you discover your purpose, and you're taking actions to help you bring that purpose into the world, you're a step closer to experiencing the American Dream.

Notes

1. Phillip Brickman, Dan Coates, and Ronnie Janoff-Bulman, "Lottery Winners and Accident Victims: Is Happiness Relative?" *Journal of Personality and Social Psychology* 36, no. 8 (August 1978): 917–27. https://www.researchgate.net/publication/22451114_Lottery_Winners_and_Accident_Victims_Is_Happiness_Relative#:~:text=Study%201%20compared%20a%20sample,a%20series%20of%20mundane%20events.
2. Zameena Mejia, "Harvard Researchers Say This Mental Shift Will Help You Live a Longer, Healthier Life," CNBC. https://www.cnbc.com/2017/11/21/harvard-researchers-say-a-purpose-leads-to-longer-healthier-life.html.
3. Eric S. Kim, Jennifer K. Sun, Nansook Park, Laura D. Kubzansky, and Christopher Peterson, "Purpose in Life and Reduced Risk of Myocardial Infarction Among Older U.S. Adults with Coronary Heart Disease: A Two-Year Follow-Up," *Journal of Behavioral Medicine* 36, no. 2 (April 2013): 124–33. https://pubmed.ncbi.nlm.nih.gov/22359156/.
4. Eric S. Kim, Ichiro Kawachi, Chen Ying, and Laura D. Kubzansky, "Association Between Purpose in Life and Objective Measures of Physical Function in Older Adults," *JAMA Psychiatry* 74, no. 10 (2017): 1039–45. https://jamanetwork.com/journals/jamapsychiatry/article-abstract/2648692.
5. Anthony L. Burrow, Patrick L. Hill, and Racher Sumner, "Leveling Mountains: Purpose Attenuates Links Between Perceptions of Effort and Steepness," *Personality and Social Psychology Bulletin* 42, no. 1 (January 2016): 94–103. https://pubmed.ncbi.nlm.nih.gov/26563209/.

PART

Discover What Stands Between You and Your American Dream

"When someone shows you who they are, believe them the first time."
—*Maya Angelou*

I went to work for my father at age 22, and by 25, I realized that something was terribly wrong with the investing industry. What I had learned wasn't adding up. The broker-dealer my father and I worked for told me what to sell, so I put my clients into those investment vehicles, only to find that those products consistently underperformed. I took the advice of the so-called experts because I assumed they knew more than me, but few of their predictions materialized. I'd have my clients work with managers who had beaten the market for years, only for those same people to no longer be listed among the top half of managers the following year. Clients weren't getting the returns they anticipated. Worse, they weren't even beating the market. I then had the thankless job of telling them that the investments I put them in didn't produce.

That didn't go well. They looked at me as if I was a buffoon, and I felt like one. I was doing the same exact thing with my own money and getting the same poor results. If my broker-dealer could not tell me in advance what managers were going to perform well, what good was I? I soon began to feel like I was pushing a narrative that was seductive and persuasive, but blatantly false.

My clients had a right to complain, but I couldn't pinpoint where I went wrong. It was my job to find the managers with the best track record in the hopes that they would continue their winning streak into the future. Why did clients even need me if I couldn't help them beat the market? Could they do just as well on their own? Those unanswered questions weighed heavily on me and kept me up at night. As a result, I couldn't eat, and my health suffered. I needed to get answers. Fast.

I went to my broker-dealer's annual conference, which was paid for and financed by the same investment companies that were failing my clients. These conferences were always in a swanky, upscale resort, typically in the tropics. Of course, if you moved enough product, they paid for your trip. All it really cost you was your soul. No one was doing much critical thinking. It was the company's intention to pump past performance and subdue any complaints or higher-level analysis of what we were doing to investors.

There was plenty of food and booze to keep the troops lubricated and happy. And they were lubricated. On the first day, as all 500 advisors mingled around the pool and listened to the band, I watched an older woman who had a few too many drinks wobble and fall right into the deep end of the pool. I kept waiting for her to come up, but she sank right to the bottom as if she were wearing lead shoes. *Somebody has to jump in and save her,* I thought, yet nobody moved. I had a brand-new suit that I couldn't really afford, but I knew that I couldn't let this woman drown because of my suit. I was a certified lifeguard and on the swim team in high school, so I dove in, and pulled her up to the surface. Two people helped her out of the water, but that was it. Everyone else was too self-absorbed, stunned, or drunk to help this woman who was in trouble. Not even the people who worked for the company and were responsible for the safety of those at the event did anything. It didn't occur to me until later, but that was a perfect representation of how they treated investors. They didn't care if you were drowning or losing money; they were only looking out for themselves.

On day two, the president of the huge brokerage firm took the stage in front of the entire group and told us that we could win a trip to the Bahamas if we sold more next year. That's when I stood up, raised my hand, and said, "I have a problem." I was the youngest guy there, and he looked at me as if I were a punk. "I've been doing exactly what you've told me to do, and my clients are getting gutted year after year. This is not working."

"I don't understand the problem." He was playing dumb, trying to draw me out.

"I don't have any confidence in what I'm selling."

"Well, when they complain, sell them something else and make a new commission. We offer enough products that something is always raging. That's how we play the game."

"I don't feel like that's the best thing for my investors."

"You don't have to do the 'best' thing. You just need to sell them what they are qualified to buy and what they want. Sell them a new product, get another commission. If that doesn't work, go out and find new investors."

That was one of those conversations I would later play over and over in my mind for years. I've thought of 100 things I could or should have said, but in that moment, I was speechless, so I sat back down. That's when I had a horrifying realization. My broker-dealer didn't care if what we were selling worked or helped people. It was all about selling products to make money. Sell 'em, churn 'em, burn 'em, and get 'em to buy new stuff. The model was driven almost entirely by commissions, and everyone in the industry was hooked. It didn't matter if it was annuities, mutual funds, or real estate; if it paid a commission, we sold it. So much of what I had bought into and believed since I was a young kid was starting to fall apart. I wasn't an entrepreneur. I was just a salesperson with a high-paying job. Nothing against salespeople, but I wasn't creating a business, and I certainly wasn't helping people. That's what

I had wanted to do from the very beginning, so something had to change.

Six months later, I made the long trip from Cincinnati to Concord, New Hampshire, to visit the president in his home office. He had me wait alone at an intimidating boardroom table surrounded by 20 chairs. When he finally walked into the room after a half-hour, I presented a plan for how I wanted to invest my clients' money and eliminate commissions. I proposed working with other advisors and helping them do the same. As a registered representative of his company, I would need his permission and blessing to launch my new plan.

I didn't get far into my presentation before he stopped me. "Are you kidding? Fidelity, Templeton, and American Funds pay us to promote their products. If I let you do what you want to do, we'll lose a ton of money and infuriate our biggest fund providers. You work for me, so I'll determine what you offer your clients. Besides, you make too much money in commissions to go out on your own and start over." Now it felt like I was drowning as he continued, "You might not realize it, but you are trapped in a pair of golden handcuffs. Go down this road with this half-baked dream and you will lose it all."

"You don't know anything about me or what I'm prepared to do," I told him.

"I know you are just a hillbilly, like your dad. You will never make it without me and this company backing you. There are a million kids like you who think they found a better way. The world chews them up and spits them out. You won't be any different."

I was ready for this moment. I calmly reached into my briefcase, pulled out my resignation letter, and slid it across the table. "I'm done. Keep your commissions." I stood up and left him sitting there at the table. I never felt more alive than I did when walking out of that room.

So, at the age of 27, I left my commission-based business, and together with my father, we became our own Registered Investment Advisors (RIAs). I quickly ran up $30,000 of credit card debt and cashed out my retirement funds to start the company. We had been working out of a converted car wash and put enough money together for a nicer office space. It was a tremendous risk, but it got me out of the commission business and over to the fee-based advisory side. I was now a fiduciary, and I felt like I could better serve my client, not the broker-dealer, which allowed me to sleep at night. I could look at myself in the mirror when I got up in the morning. And I knew I was onto something. Commissions had to go, but I was still relying on a very flawed system by using actively managed mutual funds. That would soon change.

The next piece of the puzzle fell into place when I attended a seminar in San Francisco sponsored by Charles Schwab. There was a debate between Donald Yacktman, a five-star manager with an impressive 15-year track record, and a little-known economist and academic named Rex Sinquefield. The subject was simple: do markets work? If markets fail to price information into the cost of stocks, then good managers should be able to take advantage of that to beat the market and add value to their clients' portfolios. However, if markets are efficient or in other words if they worked properly, then it is theoretically impossible to predict and beat them over time. That would mean that those who succeeded in the past were lucky, not skillful. Furthermore, if they were lucky and not skilled, they would have only a small chance of repeating their previous market-crushing performance going forward.

I had never heard of Sinquefield, but I was a big fan of Yacktman and couldn't wait to hear what he had to say. However, once he got on stage, he started rambling, and after 20 minutes, he still couldn't explain how he managed money. It made no sense. His process was a black box filled with a hodgepodge of investing jargon but no understandable approach. Then, Sinquefield got up on stage and said something that I will never forget. Something that changed the trajectory of my life:

It's my contention that active management does not make sense theoretically and isn't justified empirically. Other than that, it's okay. But it's easy to understand the allure and seductive power of active management. After all, it's exciting. It's fun to dip and dart, pick stocks, time markets, and get paid high fees, and to do it all with other people's money.

It was a gut punch. Yup, that described me perfectly. That speech confirmed what I had started to believe—money managers were speculating and gambling with their clients' money. Some got lucky and beat the market, but most didn't because stock market prices were random. I felt like a snake. How could I speculate and gamble with clients' money? I looked around the room to see who else came to the same realization, but I was all alone. Nobody else seemed remotely concerned or even wanted to admit the possibility that the way we were investing clients' money was seriously flawed.

It wasn't only my clients' money that I was gambling with; I was doing the same exact thing with my own money. I didn't resist the truth or try to rationalize my previous behavior. In the futile fight to pick the best active managers and beat the market, I fully surrendered. I was humbled to finally understand what was going on. It only took an hour, but after that debate, I knew that there was no way that I would return to stock picking, market timing, or track record investing. I could never offer my clients actively managed mutual funds again. This idea that a manager could pick the best stocks, get clients in and out of the market at the right time, tell them what sectors to be in, and give them a forecast about the future so they can beat the market was based on a fundamental lie. Managers had no idea. No clue. None.

The Investing Industry Is Broken

"One's dignity may be assaulted, vandalized, and crudely mocked, but it can never be taken away unless it is surrendered."

—*Michael J. Fox*

"How do you know if the person you trust with your money is prudently investing it for you or looking out for their own best interests?"

W hen my father was a kid growing up in West Virginia, his family could only afford to buy him one pair of shoes a year. They couldn't afford tennis shoes, so every fall, right before the start of school, his sister would take him to buy the same pair of plain black shoes. He had to make those shoes last the entire year, which never happened because he either destroyed them or outgrew them. Either way, he usually had to go barefoot by the time summer rolled around.

My father's raggedy shoes did not go unnoticed by the other kids at school. The gym coach wasn't a big fan of my father's foot-wear either because they would scuff up the floor when the kids played dodgeball. So, the coach made him take off his shoes and play in his socks. If you've ever tried to run around on a hardwood floor in socks, you know you don't make it far. You might as well be walking on ice—you slide all over the place. My dad was small and poor. This made him a literal target when playing dodgeball. A group of older kids would take those great big red rubber balls and whale on him. He'd come home with welts all over his body.

The bullying only got worse as time went on. Eventually, my father reached his breaking point and got angry. He had to stand up for himself, so he promised that he would fight back, no matter what it took—it didn't matter how big the other kid or how many of them there were.

As soon as he started fighting back, something amazing happened. The bullies stopped picking on him. He realized that deep down bullies are cowards. They don't pick on other kids because they're brave; they pick on the other kids who don't have the power to fight back. The only thing that a bully understands is strength, and if you have the strength and courage to stand up for yourself, the bullies will back down.

Years later, when I was a young kid, I started getting bullied, too. I felt like a coward, and that kept me up at night. So, when my father came home early from work one day to find me sulking on the front steps after being bullied by a group of older kids at school, his response wasn't to go to the principal or the police. Instead, he told me to pick out the biggest kid in the group and punch him square in the nose. That got me excited. "Then, I'll beat him up, right, dad?"

"Oh no. He'll beat the crap out of you," he said. "But at least you'll have a chance. It won't last forever. Despite what it feels like at the time, it will be over fast, but you have to get it over with. Then, you'll be able to get your pride back."

"What about the other guys? Won't they jump in?"

"They're cowards. They won't do anything."

My dad was so certain about it that I trusted him. So, the next time that group of bullies targeted me, I took his advice. I did exactly what he said, and he was right. Everything played out like he said it would. I got the crap beat out of me, and it was over in less than three minutes, but guess what? The bullying stopped.

More important, I went from feeling like a coward to feeling empowered. I walked around with my head up and could finally sleep at night. It wasn't the last time I was bullied, but from then on, I had the courage to stand up to the bully. That's why I grew up hating bullies, which is why it hurt me so much later in life when I realized that I had become one as an adult.

When I began working with my dad in the 1980s, the movie *Wall Street* had just come out. This was the era of slicked-back hair and suspenders. At the time, that was me, and I have the pictures to prove it. I may have only been in my 20s, but I had been watching my dad my entire life. I saw him sell insurance in West Virginia before we moved to Cincinnati, where he opened his own financial planning company. He sold stocks, bonds, mutual funds, gold, and everything considered an investment. I was doing what I had learned and thought I was helping my clients, but when I attended that conference in San Francisco and heard that debate between Rex Sinquefield and Donald Yacktman, I realized that I was the bully. I was the one promoting speculating and gambling my client's money, their futures, and their dreams. It was never my intention. I thought I was doing good, but ignorance is not an excuse.

Once I made that realization, my eyes were wide open. I could see the traditional financial industry for what it was—a disgusting and unethical field. Many in the industry keep their clients in the dark and put them in investments that are not in their clients' best interest. They certainly don't try to educate their clients on their options or train them on how to invest. They put them in a model and then disappear. They are bullies and do what's in their own best interest. Learning that I was one of those bullies permanently changed something inside me. It made me commit to doing things differently. Since then, no amount of money has tempted me to speculate and gamble with my clients' money. Unfortunately, I can't say the same for the rest of the industry. Consider this chapter a peek behind the curtain provided by someone who used to be on the other side and was part of the problem. Get ready for some inside baseball.

How Wall Street Really Works

Bullies come in all shapes and sizes, but one thing they all have in common is that they habitually seek to harm those they perceive as vulnerable. They exist everywhere, both in the real world and the virtual world, online. You may rely on them to manage your money. When it comes to identifying the Wall Street bullies, there are three types to look out for.

Con Artists

This type of bully cheats others out of their money by tricking them, gaining their trust, and persuading them to believe something that's not true. You might remember some of these names: Reed Slatkin (orchestrated a $593 million Ponzi scheme), Thomas Peters ($3.6 billion Ponzi scheme), and Allen Stanford ($7.2 billion Ponzi scheme). And then there is the person who ran the greatest Ponzi scheme of all, Bernie Madoff ($65 billion).

Bernie Madoff did not execute a single trade for his advisor clients ever. What he did was deposit his investors' funds into a bank account so he could pay off old customers with the funds from new customers. He provided his clients with false account statements. The $50 billion in returns he documented were pure fiction. It was a classic pyramid scheme.

I know what you're thinking. *I'm sophisticated. I'll never fall for something like that.* Really? Because many very sophisticated people did fall for it, such as actress Uma Thurman and her partner Arpad Busson, a French hedge fund manager. Together, they lost $270 million to Madoff. Walter Noel, another hedge fund manager, lost $7.3 billion. These aren't just guys off the street. These are smart and successful people in the financial world who were duped. Add celebrity investors like Steven Spielberg, Kevin Bacon, John Malkovich, and Larry King to that list.

I've learned over the years that it's often the most sophisticated people who are the easiest to con, and the fallout can be devastating. Some people assume that because many of these clients are already wealthy, they aren't affected the way the average person is, but that is far from the case. French businessman and money manager René-Thierry Magon de la Villehuchet lost $1.4 billion of his and his clients' money to Madoff. The shame and guilt were so overwhelming that René committed suicide. Bernie Madoff's son, Mark Madoff, also took his own life. And he had absolutely no knowledge of what his father was doing. He was just an innocent victim whose life was ruined by his father's crimes. Bernie Madoff left a trail of destruction and ruined lives in his wake.

Madoff executed a "soft con" where he promised people a 10% rate of return (which is roughly what the S&P has earned historically) but without the volatility. He didn't promise massive quick returns, like most Ponzi schemes today. That was his pitch, and because people were greedy, they fell for it. He was smart and seductive enough to sound reasonable, but there is no free lunch.

Although Madoff orchestrated an illegal Ponzi scheme, there are even more advisors who engage in unethical behavior. They dupe people into investing money with the hopes of making huge returns. Con artists can take advantage of investors in all kinds of legal ways. This has been true throughout history, and it's not going away anytime soon. The general rule of thumb has become a cliché, but it's still true: if someone comes to you with an offer that seems too good to be true, it most likely is, so run as fast as you can in the other direction. Always do your due diligence and err on the side of caution before investing your money. It's worth me repeating it: there is no free lunch.

Prognosticators

These are the people who tell you, usually with the utmost confidence and certainty, exactly what's going to happen in the future. This is common with investing, the economy, and political events. These people paint themselves as experts and try to convince you that they know how you should invest your money. This should sound familiar because it happens on every cable news show. People love making predictions. But how often do any of those predictions come true?

In 1999, Harry S. Dent Jr. published *The Roaring 2000s: Building the Wealth and Lifestyle You Desire in the Greatest Boom in History*. In it, he predicted—you guessed it—a giant stock market boom in the early 2000s. What happened? The S&P500 returns between 2000 and 2009 on US large stocks were −0.95% per year. If you listened to Harry, you lost money. But that didn't stop him from making further predictions. In his 2009 book, *The Great Depression Ahead: Strategies for a World Turned Upside Down*, he predicted—you guessed it again—a depression to follow the 2009 real estate crash. That didn't pan out either. He followed that up with his 2011 book, *The Great Crash Ahead* and his 2014 book, *The Demographic Cliff: How to Survive and Prosper During the Great Depression Ahead*, and in 2017, *The Sale of a Lifetime: How the Great Bubble Burst of 2017–2020 Can Make You Rich*. You can see how these predictions panned out in Figure 5.1.

In all fairness, one of Harry's predictions eventually came true in 2020. Hey, even a broken clock is right twice a day. People do get it right occasionally. But they tend to get it wrong much more frequently. The reality is that nobody can predict the market because nobody can predict the future. Anyone who tells you differently is either lying or delusional.

Gurus

You know these people. They are the acknowledged leaders and chief proponents in a particular field who possess the supposed expertise required to guide you toward a successful future. Guru propaganda runs rampant through the financial industry, and you'll frequently see outrageous claims about their accomplishments.

One such guru is Arnold Van Den Berg, who, according to the April 2003 issue of *Financial Advisor*, beat the market over 29 years. That may be true, but statistically speaking, someone will always get lucky. What nobody can tell you is who will beat the market for the next 29 years, because nobody knows that. Gurus rely entirely on their track record, and they only appear to be geniuses in hindsight. This article might as well be titled, "Monkeys Throwing Darts over a 29-Year Period Have Beaten the Market." That's not a joke or hyperbole.

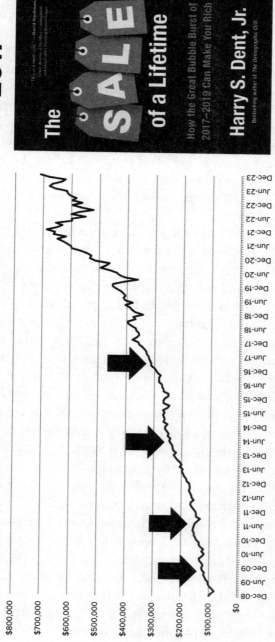

S&P 500 INDEX GROWTH OF $100K GROSS OF FEES (1/2009 – 12/2023)

Figure 5.1   How Often Does Harry Get It Right?

Note: PAST PERFORMANCE IS NO GUARANTEE OF FUTURE RESULTS. This information is for educational purposes only and should not be used as investment advice. Investors cannot invest in a market index directly, and the performance of an index does not represent any actual transactions. The performance of an index does not include the deduction of various fees and expenses which would lower returns. All investing involves risks and costs. Even a long-term investment approach cannot guarantee a profit. No investment strategy (including asset allocation and diversification strategies) can ensure peace of mind, guarantee profit, or protect against loss. This figure does not reflect actual performance of any managed portfolio and no representation is made that your portfolio would achieve similar results. The S&P 500 Index and associated risks are discussed more fully in the endnotes. See Appendix II for additional information.

Source: Harry S. Dent Jr., *The Sale of a Lifetime* (New York: Simon & Schuster 2017).

In 1973, Princeton University professor Burton Malkiel wrote in his book, *A Random Walk Down Wall Street*,[1] "A blindfolded monkey throwing darts at a newspaper's financial pages could select a portfolio that would do just as well as one carefully selected by experts." Twenty-five years later, Rob Arnott, CEO of Research Affiliates, put that to the test and discovered that the monkeys did better than the experts and the market.[2] That won't keep Fidelity, Vanguard, American Funds, Morgan Stanley, or Merrill Lynch from telling you they can still pick the best stocks. And rarely do clients ever ask for scientific evidence proving how they can do it. We just take it in good faith. *These are all big financial companies. They must know something, right?*

Have you ever watched Jim Cramer on television? He's not licensed or regulated, but he's made a very good career talking about picking stocks as an entertainer. CNBC even gave him his own show, *Mad Money with Jim Cramer*, where he provides stock picks and predictions. The Action Alerts PLUS portfolio was based on Cramer's picks. As of December 2020, after 17 years of fund performance, the returns of the Action Alert PLUS portfolio were 189.97%. That sounds impressive until you consider the returns on the S&P 500 were 349.57% during that same period. If you had kept your money in an S&P 500 index fund, your return would have been 160% higher. And get a load of this: The Action Alerts PLUS portfolio has only narrowly beaten the market seven times between 2002 and 2021.[3] It typically gets destroyed, and as of September 2021, Jim Cramer is no longer affiliated with the portfolio. And they call this guy a guru and give him his own television show. But you don't need to be a bona fide expert anymore to attain guru-like status.

Did you watch the 2022 Super Bowl between the Los Angeles Rams and Cincinnati Bengals? It was also called the Crypto Bowl because practically every other commercial had something to do with cryptocurrency. You had Tom Brady, Stephen Curry, Lebron James, and Larry David, all out there shilling for crypto. Matt Damon told us, "Fortune favors the brave!" For a while, all we heard about was crypto. How did that turn out? Since then, Sam Bankman-Fried, disgraced founder of crypto exchange FTX, lost $8.9 billion of his clients' money before being found guilty of wire fraud, conspiracy, and money laundering. In 2024, he was sentenced to 25 years in prison. More than $2 trillion has been wiped off the value of cryptocurrencies since its peak in November 2021.[4] The problem with cryptocurrencies is that you're not buying a company, intellectual property, human capital, or anything tangible, so there is no value. It's not even a real currency because it's not stable. I believe it has no place in a prudent portfolio.

I recently discovered a quote by Andrew Vachss that I love: "Life is a fight, but not everyone is a fighter; otherwise, bullies would be an endangered species." And all of these gurus are bullies. Jim Cramer—bully. Tom Brady is an amazing athlete and I admire him in many ways, unfortunately he has hurt a lot of people in his support of FTX. It doesn't matter that he isn't purposefully trying to hurt you with the products he promotes, but just as in my case, ignorance is not an excuse. Tom Brady can claim plausible deniability and say he's not an expert, but he was still paid millions to get people to speculate and gamble with their money. You can be a bully on purpose, or you can be a bully out of ignorance. Ultimately, it makes no difference to the person getting bullied and losing money.

How the Experts Get It Wrong

If you ever find yourself being seduced by a prediction, do something that almost nobody ever does. Go back and look at how some previous predictions panned out. It will scare the living heck out of you.

In the July 19, 1993, Morgan Stanley's Barton Biggs encouraged investors to sell US stocks. He said, "I want to get far away from Bill and Hillary. The President is a negative for the U.S. market."

I hope too many people didn't listen to Barton because the S&P 500 returns of US large stocks from July 1993 to December 1999 was 22.5% per year. That's a total return of 273%. Oops! Did Mr. Biggs issue an apology in case you took his advice and missed out on those massive returns? If he did, I could not find it. Do magazines learn their lesson and realize that these types of predictions can hurt people? Ha!

On August 14, 2000, the headline on the cover of *Fortune* magazine read, "Retire Rich," and inside they listed "Ten Stocks to Last
the Decade"[5] (see Figure 5.2). In the article, they said, "[We've]
identified four sweeping trends that appear certain to transform
the way we work and interact . . . [and then] sought out some of the
top stock pickers in the country."

Figure 5.2   Fortune (August 14, 2000)
Note: See Appendix II for additional information.

In this context, "appears certain" is a euphemism for "guarantee." So, how do you think these 10 stocks did over the next decade? Well, in the 10-year period between July 19, 2000, and July
16, 2010, you would have LOST 47% of your investment if you had
purchased these stocks, based on their share prices. Do you think
Fortune acknowledged that? No! They hope you forget they wrote
the article.

Just one month earlier, on July 24, 2000, *Fortune* ran a cover story
with the headline, "Let Them Make You Rich"[6] (see Figure 5.3).
The article featured a panel of "all-stars" who made 29 stock picks.

How do you think those performed over the next five years? Well, 16 were no longer trading under their original name or ticker due to mergers and acquisitions. Here were the top four on the list from the price on July 5, 2000 to the price on July 5, 2005:

1. eBay ($5.32 to $14.12)
2. Target ($28.84 to $56.04)
3. Best Buy ($27.22 to $47.17)
4. Goldman Sachs ($95.94 to $103.18)

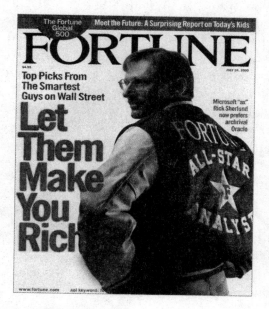

Figure 5.3   Fortune (July 24, 2000)
Note: See Appendix II for additional information.

Okay, not terrible. How about the others?

- Citigroup went from $474.38 down to $464.60.
- Priceline went from $222.75 down to $23.26.
- Estee Lauder went from $24.50 down to $19.41.
- Applied Materials went from $41.72 down to $16.27.
- Teradyne went from $73.13 down to $11.71.

- Amdocs went from $73.00 down to $27.79.
- CMGI went from $443.75 down to $19.20.
- Global Crossing went from $26 down to $0.
- WorldCom went from $45.00 down to $0.

If you invested $10,000 in each of these 13 stock picks made by the "all-stars" in July 2000, that $130,000 would be worth $102,557 five years later. Yes, you would have lost money based on market price!

What makes some of these predictions so egregious is the language used to promote these stocks. CMGI was referred to as "a screaming buy." Global Crossing was predicted to be a "$70 stock," and you were told you could pick up WorldCom "dirt cheap"—both stocks went to $0. Many people think *this can't happen to me*, but a lot of people owned these stocks.

These predictions above aren't anomalies. They happen all of the time. Let's jump ahead 20 years to see how things changed.

The December 2020/January 2021 issue of *Fortune* featured an article titled, "The 21 Best Stocks to Buy for 2021"[7] (see Figure 5.4). Did they do any better this time around?

- Average return of the "21 Best Stocks of 2021": 16.49% based on share price
- S&P 500 return: 28.71%

"THE 21 BEST STOCKS TO BUY FOR 2021"

December 2020/January 2021

JANUARY 1, 2021 – DECEMBER 31, 2021

Average Return for the 21 Stocks   16.49%
(based on share price)

S&P 500 Index Annual Return   28.71%

Figure 5.4   *Fortune* **Picks for 2021**

Note: See Appendix II for additional information.

I don't know about you, but I'll take S&P. No stock picking required. But how do predictions fair during a down year?

Exactly one year later, *Fortune*'s cover story read, "11 Steady-Rising Stocks to Own for 2022"[8] (see Figure 5.5). Let's assess the damage.

- Average return of the "11 steady rising stocks": −$32.46% based on share price
- S&P 500 Return: −18.11%

"11 STEADY-RISING STOCKS TO OWN FOR 2022"

December 2020/January 2022

JANUARY 1, 2022 – DECEMBER 31, 2022

Average Return for the 11 Stocks   −32.46%
(based on share price)

S&P 500 Index Annual Return   −18.11%

Figure 5.5   *Fortune* Picks for 2022
Note: See Appendix II for additional information.

Even when it loses money, I'll take the S&P instead of relying on stock picking, market timing, and track record investing. In case you missed the point, sometimes you might get lucky and pick stocks with high returns, but not even the professionals can tell you what they are in advance. That is why it is gambling. If they really knew, with anything like certainty, what stocks were going to soar into the stratosphere they wouldn't print the losers, and if they really did know how to pick 'em they wouldn't tell you for free.

In 2023, the S&P experienced a 26.29% gain. Nobody saw that coming. Almost all of the prognosticators and gurus were bearish on the market. Many even predicted the S&P would decline for the first time since 1999, but that didn't happen, and it didn't stop those same people from revealing their 2024 predictions. This trend is not going to change any time soon. People will always be seduced by predictions because we so desperately want them to be true. Predictions are happening every day all around us. They sound sexy and enticing, but which ones should you believe? The simple answer is none. We all want to think that we have a scoop or insider info that will get us just a bit ahead of the next person. In reality, it's all speculation and gambling. A lot of time, money, and thought goes into selling you on the illusion that these bullies know the future and their predictions will pan out.

How the Illusion Is Crafted

The investing industry is filled with illusionists. I'm talking about the con artists, prognosticators, and gurus. I'm also talking about the media and all the Vanguards, Fidelitys, and BlackRock—any company that pretends to have brilliant people who can pick the best stocks in advance and pass on all the extra profits to you. They are all part of the same rotten system and embedded in every facet of the industry. It takes time, effort, and energy to create the illusion, and the industry has invested billions. Just like illusionist David Copperfield made the Statue of Liberty disappear on national television in the 1980s, those in the investing industry have been trained and are always honing their methods to get your money. It starts by scaring the absolute hell out of you.

Preying on Fear

If the illusionists can scare you, they can manipulate you and profit by destroying your American Dream. They prey on fear, and there is no shortage of things to be afraid of in America today. Fear comes in many forms but can also be unique to you. Think of your three biggest fears. If you need help, here is a common list:

- Being alone
- Losing a loved one
- Death
- Wasting your life
- Living a boring life
- Your spouse leaving you

- Losing your job
- The IRS
- Poor health
- Hurt by crime and violence
- The government
- Something negative happening to your child
- Living an unfulfilled life
- Becoming a burden for your children
- The cost of living
- Running out of money in retirement
- Wasting time
- Nuclear war

I pulled these answers from a recent workshop. Did you notice there are still some money demons included in there? That's just another reminder of how money weaves its way through every facet of our lives, including our fears. The media knows this better than anyone, so they pump fear and uncertainty into the culture 24/7. The investing industry uses this to make stock picking, market timing, and track record investing much more appealing.

If you boil all these fears down to a single statement, it would be the fear of not "making it" in life. Here's a secret: none of us will "make it" in life. We all end up in the same place. As a minister once told me, "The statistics on death are impressive." So, the question shouldn't be whether or not you're going to make it. The question should be what you will do between now and . . . whatever time you check out. What are you going to do with the time you have left? This is where the American Dream and purpose come in, but it becomes significantly easier to succumb to fear if you don't have purpose.

The Law of Large Numbers

Once the illusionists have you scared, they lure you in with hope. Everybody wants a way out of difficult times. Everybody wants to get a little bit ahead, just a cushion to give them some breathing room. And nobody is going to turn down easy money, so that's what the industry promises. They do that by convincing their clients and the general public that they can predict the future and pick the right stocks. Nobody can predict all of the things in the future that will predict stock prices, but the industry perpetuates the illusion through several tricks; one of the most effective being the law of large numbers.

Pretend we have our own mutual fund, and you're the stock picker. You know you can't predict the future, so you must devise a way to fool the public into thinking you can. Here's how it works. Think of a stadium filled with 100,000 people. Every single one of those people will stand up and flip a coin. Probability tells us that half will flip heads and the other half tails. The half who flip tails sit down, leaving 50,000 people still standing. Those people represent the money managers who beat the market in a given year. The same is true of you. Half of the investing public could beat the market without skill. That's how the monkeys beat the managers. However, magazines and cable news always say something like, "This is the person who will keep doing it for the next 30 years!" But in reality, they have no clue.

Let's have the 50,000 people left standing to flip the coin again. The odds say that 25,000 people will flip heads twice in a row. If we keep going, 12,500 flip heads three times in a row and 6,250 four times in a row. If you were playing along at home and flipped heads four times in a row, congratulations! This is all you need to be a five-star mutual fund manager. You're crushing it. You'd be the guru on TV that everyone says is brilliant, even though what got you to this point is total luck.

If we keep going, 3,125 people will flip heads five times in a row, and 1,562 will flip heads six times in a row. You get the idea. Do the math; there is just over a 1% chance to "beat the market" by flipping heads six times in a row. Those odds don't sound appealing, but because we started with a sample size of 100,000 people, that 1% comes to 1,562 people. That kind of sounds like a lot now, doesn't it? We could have kept going, and in that crowd of 100,000 people, there would have been some people who got lucky 8, 9, 10 times in a row. If you have a big enough sample size, many people will get lucky, and those people will try to palm that luck off as skill. They might even believe it themselves. The most convincing hucksters and snake oil salespeople believe their own lies, and that's what makes them so dangerous. It doesn't matter if they're malevolent or incompetent, it still hurts you.

If this is your mutual fund company, and you get lucky 10 times in a row, of course, you say it's because you have a brilliant system. But it's really just luck. When it works, it's a good business model. It's terrible for investors, but who cares about them if it's a good business model? That's what large mutual funds and brokerage companies do. They have a big room full of people basically flipping coins with their thousands of funds. And guess how many total mutual funds are available in the United States? According to Morningstar, as of December 31, 2023, there were 20,424. What is the total number of stocks you can invest in? According to the same source there are 20,819. That means there are many more mutual funds than there are individual stocks. If that doesn't prove to you that something is deeply wrong with the system, then I don't know what will.

The industry is manipulating the public and creating the illusion that they stock pick and market time to beat the market. Still not convinced? Tell me this: if Fidelity or any other mutual fund company knew who the best manager was and that manager could consistently beat the market, how many funds would they need? Simple logic tells us they'd only need one. It would be a big fund, and they wouldn't have to do much marketing. And if that manager overseeing that one fund knew the best stock, why would they bother with any other stocks if they knew they wouldn't do as well? Logically, they would put all of their money in the best stock. Advisors and money managers either don't know this or don't care and are in on the illusion.

Here's another question: if there was a prognosticator or a guru who could pick the best stock and knew how to get in and out of the market at the right time to maximize returns, why would that person ever tell you? They'd be making 30% to 40% a year. They'd never lose money. They'd have billions of dollars—more money than Zuckerberg, Bezos, and Musk combined. They would be the richest person on the face of the earth. Why go on cable news and give that information to the public for free?

This illusion is one of the greatest scams perpetuated against the American public, and unsuspecting investors continue to be manipulated. There is a war being waged for your mind. It starts by instilling fear and controlling how you think. Once they control how you think, they can then influence your behavior. And if they can influence your behavior, they can get your money. That's what they want. The best way to protect yourself is to identify and avoid the bullies.

Is Your Manager a Bully?

Let's revisit the question from the beginning of the chapter. *How do you know if the person you trust with your money is prudently investing it for you or looking out for their own best interests?*

What makes an effective bully difficult to recognize is that they are likable. They are charming and persuasive. They appear trustworthy. People like Jim Cramer. They liked Bernie Madoff. They absolutely love Tom Brady. That's why you can't judge a bully by their likability. You must judge a bully by their behavior. If you're worried that your current advisor fits into this category, first determine if they are doing any of the following three things. Fair warning, you might start to experience cognitive dissonance.

1. **Stock picking.** This is when you or an advisor use a systemic form of analysis to conclude that a particular stock will make a good investment and should be added to your portfolio. If you currently have individual stocks in your portfolio, you don't have to look hard for evidence that you've engaged in this behavior. You're doing it now. If your advisor encourages you to stock pick, they are part of the problem.
2. **Market timing.** This is any attempt to alter or change a mix of assets based on a prediction or forecast about the future. If your advisor encourages or facilitates you getting in and out of the market at specific times, they are part of the problem.

3. **Track record investing.** This involves relying on the performance history of a sector or manager to determine the best investment for the future. If an advisor sells you mutual funds or other investments that have done well in the last three or four years without properly educating or training you, they are part of the problem.

These behaviors often include endless sales presentations and a lack of real communication. These advisors will show you data that makes it look like they've helped real people like yourself beat the market. One thing about the industry is that they often avoid showing you their real returns. They'll show you different mutual funds but won't say, "This is our company as a whole, and this is exactly what our investors have made. We've been recommending this for the past 30 years, and this is how our clients have done based on our recommendations." In my experience, the returns these companies flaunt are seldom audited, and rarely include all of their clients. Instead, they take the path of least resistance. They pick three or four investment vehicles that have performed the best (or ones they just got lucky on) over the past three years. They put together a pie chart and say, "If you had your money with me, you would have been in these investments and made 25% over the last three years."

I'm always asked, "Don't advisors make more money if they help their clients make more money?" The answer is no, not all the time they don't. It's far easier to manipulate you and just show you what investment vehicle had the highest return over the last 1, 3, 5, or 10 years and seduce you into investing in that versus spending days, weeks, or months educating you on the academic research.

There is a good chance that many of you will recognize that you are working with and trusting your money to a bully who is part of the problem. What's crazy is what often happens after people make this realization.

I've been coaching for a long time, and many of you reading this book will recognize these behaviors and get mad at your advisor.

You will put your foot down and promise you won't work with that person ever again. You will take this seriously, learn about the academic research because you want to make prudent decisions. Then, out of the blue, you'll call up that same advisor and ask what they think about everything you just learned in this book. It sounds absurd, but it happens all the time. That's why I've come up with a simple rule:

Never go back to the person who created the problem and ask them to solve it, especially if they profit from keeping you in the problem.

Remember, many investing advisors, financial planners, and wealth managers are largely commission-driven. They earn a profit by getting you to make certain investments; they don't need those investments to pan out. That means they have a conflict of interest when managing your money. There are a variety of reasons investors overlook this. It could be avoidance. People want to avoid confrontation, and they want to avoid taking responsibility for their actions because, if anything goes wrong in the future, they can blame the advisor. It might be people-pleasing. Maybe an advisor has been working with their family for decades, and they feel guilty moving their money. These reasons are counterproductive and can be downright destructive when investing. Still, there is one reason for continuing to work with a toxic advisor that is more seductive and dangerous than the others.

The Friend Filter

I've lost track of how many times a client has called me up and said something like, "Mark, you know, I've been friends with my advisor for 20 years, and I'm afraid that I will hurt their feelings and ruin our relationship if I don't invest with them anymore." I call this the friend filter, and it sounds crazy, but people do it all the time. They like their advisor and feel that they are friends with their advisor, so they don't want to hurt their feelings.

If you ever catch yourself thinking like this, stop and ask yourself a question: If you're going to board a plane and fly halfway around the world, do you want a likable pilot who you think is your friend, or do you want a pilot who can confidently get you through a horrific storm and land safely at your destination on time? What if you're about to undergo heart surgery? Do you want the surgeon you can have a beer with or the most competent surgeon you feel can do the best work? The person who manages your money has a major influence on your purpose, your American Dream, your future, and the future of your family. Why jeopardize that by investing with someone you didn't feel was the best? Will you put your friendship with that advisor over everything you value most in this world?

I'm not putting down friendship, I believe it's one of life's greatest joys. It's a key part of the American Dream. You can't experience freedom, fulfillment, and love without friendship. My closest friends have improved my life. But I don't need my money manager to be my friend. I don't need my heart surgeon to be my friend. I don't need my pilot to be my friend. I need them to be competent people of integrity. You should value those characteristics in an advisor, not charm or charisma. Friendship is an emotion, and emotions often lead to destructive investing behavior. Unless properly harnessed, emotions are your enemy when investing. You want your investing decisions to be made from the perspective of purpose and backed up by the academic investing principles you will learn in Part III. Trust should be earned through competence, not friendship or likeability.

If you're only staying with an advisor because that person is your friend, it would make more sense to prudently invest your money on your own and mail that advisor a $10,000 check for being your friend. When you break it down, the cost of doing business with that advisor might be even higher because, if they're speculating and gambling with your money, you're risking the destruction of your portfolio.

What makes the friend filter so tragic is that any true friendship will easily survive this transition. Friendship is not transactional. If your advisor is really your friend, they will remain your friend whether they're making money off you or not. Over the years, I've done business with many friends, and when that business runs its course or is no longer mutually beneficial, we stop working together. Never has that meant that I stop being someone's friend. We still talk to each other, go to lunch, and are there to support each other. A healthy friendship does not involve being able to make money off of each other. If it does, then it's not true friendship, and a good sign that you are being used.

Keep in mind that advisors go to seminars to learn how to bond and build relationships with their clients. It's a sales technique. In the industry it's called *bonding* and *rapport*. They study you and your behavior. You are homework to many industry professionals with hidden agendas who rely on manipulation and charisma to build trust. And it works because we naturally trust people we like and who we feel like us. It seems logical that friends won't hurt each other's feelings and will have the other person's best interest at heart. That's why charm is a weapon that can manipulate people. Trust the science, not what you believe to be friendship. There were so many manipulative techniques that my broker-dealer taught me, and the company didn't think twice about the ethics of it. When a client is unhappy, the answer is always "sell them something else to earn another commission." That's how many advisors think.

Seducing clients through friendship is what keeps the wheels of the industry turning. Once they've acquired that friendship, manipulative advisors don't try to build your confidence or clarify anything. They try to do the opposite by creating a confusion trap. That is when they hit you with so much jargon and complexity that you don't know what to think. It's part of their job to confuse you, and they use complexity as a weapon. Once you're confused, they attempt to convince you that their predictions are reasonable when, by its very nature, a prediction is anything but reasonable.

Trying to educate yourself is no easy task because you're met with information overload. Google the word *invest*, and you get approximately 5.8 billion results. If you attempted to read all of that information, it would take you more than 33,000 years—and that's if you only spent three minutes on each page. I make this point in my workshops, but every year I have to keep updating the numbers because there are more and more results. No matter what the number, nobody is living that long. There's no shortage of information out there, and there is no shortage of misinformation, either. You don't think all of those 5.8 billion results are reliable or accurate, do you? How would any normal, sane person ever learn the right thing to do?

This all works in favor of the advisors. If they can develop a friendship, confuse you, and get you thinking that they're much smarter than you are, they can keep you buying and selling. That's how they make money, and it works because investors are vulnerable. However, if you can successfully navigate this labyrinth to avoid all of the pitfalls and traps, you will learn that investing can be an incredibly rewarding process that is much more satisfying than the short-term dopamine hits that come from speculating and gambling.

Now, it's time to turn the tables. It's your turn to study the advisors, and if you discover that it's not in your best interest to keep working with that person, are you willing to stand up for yourself? Will you sever that relationship? Are you willing to remove the friend filter? You, and only you, can do that. You must, because your dreams are worth fighting for. That means taking the time to do the uncomfortable things in your best interest. You must confront your past mistakes and remove toxic advisors from your life. If you don't, you are jeopardizing your purpose, your family's future, and your American Dream. Your future depends on it. The process begins by applying one simple standard.

The Fiduciary Standard Versus the Suitability Standard

Successful investing is a key element to helping you fulfill your purpose and experience the American Dream. If you can maximize your potential as an investor and do what is prudent with the money you have to work with, you position yourself to live an extraordinary life. That's why who you listen to and who you trust with your money is extremely important. So, when it comes time to pick an advisor or switch advisors, you first want to ensure they are a fiduciary. Let's define our terms.

A *fiduciary* is legally obligated to do what's in the client's best interest and disclose any conflicts of interest. Fiduciaries typically earn a fee and are not paid in commissions.

The suitability standard is typically what broker-dealers follow, and it means they sell clients what they are simply "suitable" for. In most cases they *are not* legally required to do what's best for the client; they can sell a product that has been deemed to have an appropriate amount of risk for a particular client, regardless of commissions paid. Only in certain, limited circumstances are they required to follow a fiduciary standard. So, if a client has $1,000,000 to invest and can risk losing $200,000 in some product the broker-dealer put together, they can sell it to the client. Most broker-dealers are facilitators or order-takers, not fiduciaries.

What makes this so much worse is that investing firms and advisors can have a money management component of their business where they are fiduciaries, but the other side of their business involves selling products. They try to wear two different hats. Which brings up an interesting conundrum, how can you simultaneously operate under a fiduciary and suitability standard? This arrangement is rife with potential and substantial conflicts of interest.

To demonstrate this, I know of an advisor who had $100 million in clients' money under management. He was nearing retirement and began moving clients' assets paying him an approximate 1% annual fee, and dumped the money into annuities paying him about a 9% commission. In many cases there is no fire wall separating fiduciary and non-fiduciary behavior. This resulted in a massive windfall for the advisor just in time for him to enjoy his "golden years."

I've witnessed many advisors do some incredibly selfish things with their clients' money, and in most cases the clients have no idea that the advisor is no longer looking out for their best interest.

Unless a company is a fiduciary from top to bottom, they are not a real fiduciary in my opinion.

Here's why this distinction is so important: Let's say you have $1 million in a 50/50 portfolio—50% in equities and 50% in fixed income—and the market is down 50%. Because nobody knows what will happen next, and the market behaves randomly, a prudent thing to do would be to sell the fixed income (which is high relative to the stocks that are down 50%) to get back to that 50/50 mix.

The true test for any advisor is when the client comes in and says, "You're making money off the money I have invested with you. You do what I tell you to do, and I'm telling you that I want what's left of my money in fixed income. Get me out of the market, now!"

The honest fiduciary would say no to the client. "I'm not going to hurt you just to keep the account. I can't be a part of this because we should be doing exactly the opposite of what you're saying right now."

Many advisors won't do that—even fiduciaries. It's easier to tell clients what they want to hear because they can continue earning money. Just because they are legally required to act in the client's best interest doesn't mean that all of them do. Some won't put the client first if it means potentially losing their account. They tell themselves it's a necessary lie because the client is still better with them as their advisor than with someone else or on their own. However, they're still facilitating speculating and gambling. You want an advisor who would rather lose you as a client than gamble and speculate with your money just to keep you so they can continue making money off you.

Being able to recognize the illusion crafted by Wall Street bullies and having the courage to sever your ties with toxic advisors who are not looking out for your best interest is a major part of becoming 100% responsible for your future. However, that is only half the battle. As egregious as the behavior of the con artists, prognosticators, and gurus is, the biggest barrier between yourself and success is not the investing industry. The biggest barrier is yourself.

Notes

1. Burton Malkiel, *A Random Walk Down Wall Street: The Time-Tested Strategy for Successful Investing* (W. W. Norton & Company, 1973).
2. Rick Ferri, "Any Monkey Can Beat the Market," *Forbes*, December 20, 2012. https://www.forbes.com/sites/rickferri/2012/12/20/any-monkey-can-beat-the-market/?sh=595c2b8a630a.
3. Gladice Gong, "Action Alerts Plus Review: Can You Really Trust It? (2024)," Earn More Live Freely, January 1, 2024. https://www.earnmorelivefreely.com/action-alerts-plus-review/.
4. Rob Minto, "Cryptocurrencies Lose $2 Trillion in Value Since 2021 Peak," *Newsweek*, June 13, 2022. https://www.newsweek.com/cryptocurrencies-lose-trillion-value-since-peak-1715207.
5. David Rynecki, "10 Stocks to Last the Decade," *Fortune*, August 14, 2000. https://money.cnn.com/magazines/fortune/fortune_archive/2000/08/14/285599/index.htm.
6. "Let Them Make You Rich," *Fortune*, July 24, 2000.
7. Anne Sraders, Jen Wieczner, Shawn Tully, and Matthew Heimer, "The 21 Best Stocks to Buy for 2021," *Fortune*, November 20, 2020. https://fortune.com/2020/11/20/best-stocks-to-buy-for-2021-airlines-health-care-green-energy-banks-consumer-international/.
8. Anne Sraders, "11 Steady-Rising Stocks to Own for 2022," *Fortune*, December 1, 2021. https://fortune.com/2021/12/01/stocks-to-own-2022-amzn-msft-pypl-crm-pep-jnj-cmcsa-shop-tsm/.

Understanding Yourself

"As Carl Jung put it, 'in each of us there is another whom we do not know.' As Pink Floyd says, 'there is someone in my head, but it's not me.'"
— *Dr. David Eagleman*

"What cognitive biases do you succumb to in your life?"

We all like to believe that we are in control—I know I do. It's natural to think that we're like Spock from *Star Trek*, and all of the important decisions we make about our money are based on logical reasoning. Unfortunately, much of that is an illusion. When you take the time to study the brain and human behavior, you learn that control is limited. That realization is often met with cognitive dissonance. But the sooner you can fight through that, the sooner you can develop the processes and systems to help you take back some of that control.

You didn't design your brain. That was done a long time ago. You have roughly 100 billion neurons in your brain, responsible for receiving information from the outside world and transmitting commands throughout your body.[1] For a frame of reference, there are 100 billion stars in the Milky Way Galaxy. So, you have just as many neurons in your brain as stars in the galaxy. Those neurons form more than 100 trillion connections.[2] That's a lot of activity, and you don't control any of it. Not convinced? Try to pick a single nerve cell in your brain, fire it off, and have it send a message to another cell. You have no clue how to do that. The brain is largely a mystery to its user. It's like a computer without an operational manual.

There are three main variables that can cloud and distort our judgment: biases, emotions, and instincts. Biases are cognitive mistakes in thinking. Emotions are feelings. Instincts are survival mechanisms hardwired into the system. All of them work together to influence our thoughts and behaviors in ways that can be detrimental to our own self-interests. It doesn't need to be that way. It's time to learn a little bit more about what you *don't* know you *don't* know about yourself. In physics, there are problems you can't begin to solve until you consider the observer. When it comes to everything in life, including investing, you are the observer and far from reliable, so we must observe the observer.

You Have Less Control Than You Think

You're always thinking. Even when reading this book, you're thinking more about what you're reading than actually reading. Not only that, but you aren't even in control of those thoughts. Here's another experiment. Try to stop thinking. That's all you have to do. That's the whole experiment. Take out your phone, set the timer for 30 seconds, and try to stop thinking. Zen out. Do your thing. See if you can achieve complete silence in your head. Good luck!

As soon as you look back at your phone, you're thinking. If you're thinking that you're not thinking, you're thinking. If you think you're doing it, you're still thinking. It's almost impossible to stop thinking for 30 seconds, and if you can do it, you're a better person than me because I don't come close. Even masters of meditation can only keep the internal state quiet for a limited time before that inner voice returns. It's unstoppable. How can you control something you can't stop?

Our thinking often shows up in a conversation or self-talk. It's a continuous loop of endless chatter in our brains. If you were in control, you could stop that train of thought whenever you wanted, 100% of the time. That would mean you would never feel stressed or experience sleepless nights again. You could tell yourself "be happy" or "don't worry," but we know that's unrealistic because it doesn't work that way. So, how can you possibly control something you can't stop doing?

Not only are we engaging in constant self-talk, but have you noticed how little of that self-talk is positive? Consider how you talk to yourself. It's probably safe to say that your internal voice is not your biggest cheerleader. Throughout your day, is your inner monologue pepping you up and motivating you to expand your potential? What about when times are tough and life doesn't go your way? What about when you get frustrated and disappointed? Is that voice in your head always there to make you feel better?

Don't worry; nobody's self-talk is always positive, and most people are downright brutal in how they talk to themselves. There is a reason for that. Our self-talk is a way that we express our doubts and fears. It seems we are always talking to ourselves, but who is doing the talking, and who is doing the listening? It's even stranger when we begin to argue with ourselves. It reminds me of the old cartoons where the characters have an angel on one shoulder and a devil on the other. It's quite bizarre, and we don't have any control of it. If we did, we wouldn't be arguing with ourselves.

It's difficult to control your thoughts. It's also difficult to control your actions. Have you ever told yourself you'd do something and didn't end up doing it? What about the opposite: is there something you said you wouldn't do and did anyway? Whether it's eating food you shouldn't, not exercising when you said you would, not getting work done that you wanted to, or not spending the time with family that you should, this probably happens frequently. Did you ever say you wanted to get up early and then hit the snooze button as soon as the alarm went off in the morning?

This isn't a new problem or one that's unique to you. And it doesn't matter your race, religion, nationality, background, or upbringing. This has been true for everyone since the beginning of time. In a letter to the Romans, Paul said, "What I don't understand about myself is that I decide one way, but then I act another. Doing things I absolutely despise . . . I decide to do good but I don't really do it; I decide not to do bad but I do it anyway." We're constantly violating our commitments and have a voice running nonstop inside our heads, egging us on. It tries to convince us to take the easy route. That's where the "yeah, buts," "how abouts," and "what ifs" come in. That's how we rationalize not being in control of our behavior. And it's all just human nature.

Unlike our thoughts, we can temporarily conquer our behavioral impulses. You know how good it feels to resist temptation and accomplish your goals. You know what I mean if you've gone to the gym on a day when you absolutely did not want to go. But nobody's perfect. Even after exerting yourself and achieving what we perceive as control, temptation occurs, or worse, we let our foot off the gas and feel we've earned ourselves a break. That's when the desserts show up. It's not that we never have control. Sometimes we do, but it's temporary, and we can so easily revert back to our old behaviors.

In the end, if you rely on willpower alone, you will lose. It's inevitable because willpower isn't enough. Part of the problem is the way we think of willpower. People believe it's similar to building muscle, and if you exert yourself you can build up willpower. That's not how it works. Willpower better resembles a gas tank: the more of it you use, the quicker you run out. Eventually, we all hit a wall and are depleted.

This is why just giving you information about investing won't work, because knowledge alone won't help you, even when you understand the consequences. The negative effects of poor diet and lack of exercise are widely known. We know about the dangers of being overweight, yet nearly 40% of Americans are obese.[3] Many of these people understand they are doing themselves harm, but it doesn't make a lick of difference or change their behavior.

When it comes to the obesity epidemic, what role does the food industry play? Do they help people understand their options and facilitate better eating habits so they can live healthy and happy lives? I don't need to answer that for you. The food industry has zero interest in your commitment to good health. If anything, they work against it. You can't even go to Staples to buy office supplies without passing a row of candy at the checkout counter. Why? Because they are trying to tempt you into buying peanut M&Ms. And if not you, there is a good chance your kids will want something. It's just how we are hardwired as humans. The food industry knows this and preys on your weakness.

The investment industry works the same way. Today, you can go online or open an app on your phone to trade around the clock. The market doesn't even need to be open. You can stay up all night if you want, and I know many people who do. That's extremely destructive. Some people know this and still can't stop because they're addicted. And the more they do it, the more out of control they become until, eventually, that behavior becomes a part of their identity. This can tear families apart. Is the industry going to help those people? Of course not! They will enable them every chance they get. They will prey on that addiction, which is why information alone is not enough to change behavior.

Your Brain Was Not Built for Investing Success

Our brains are hardwired for survival. How do I know that? Those whose brains didn't adapt to survive didn't make it.

Our brains are busy dealing with the future that could happen. As you read this, your brain is focused on its primary job: making sure the bad future it's worried about doesn't occur. It's always on the lookout. This is where most of that negative self-talk comes into play. Your brain is trying to talk you out of doing what's challenging and convince you to take the easy path of least resistance because it's safe. As soon as your brain perceives a possible threat, even if it's imagined, it will take action to ensure you avoid it.

The cortex is the part of the brain responsible for logical reasoning, and when it perceives a threat, it shuts down, and the amygdala takes over. That's the small, almond-shaped part of the brain that keeps you in a state of panic and fear. When we react to these real or imagined threats, it's called an *amygdala hijack*. And human beings are one giant reaction. Those reactions are heightened during extreme conditions. Once that threat passes, the cortex will try to justify those actions as rational. It's why we always have a good reason for doing what we do. We can justify our bad decisions to ourselves extremely well after the fact. This is where cognitive bias comes into play.

A cognitive bias is a systemic error or distortion in thinking that affects our choices and judgment. Meir Statman, who has a PhD in behavioral finance, put it this way: "Biases have to do with cases where the intuition that generally gets us right gets us wrong." So, when you think you're making conscious decisions about investing, your subconscious is calling the shots, and you're succumbing to your cognitive biases. The subconscious brain makes up 95% of the mind.[4] That's a staggering percentage. Our rational mind, which only represents 5%, constantly tries to justify the thoughts and behaviors of our subconscious. Research has shown that the impulse for some behaviors happens faster than the conscious mind can react. Far too often, the conscious mind doesn't override the impulse.

The first step in combatting these biases is understanding what they are and how they work. If you don't understand why you're reacting in a certain way, it's much more difficult to change your actions. There are more than 175 cognitive biases[5] that can distort the way you see the world. Yes, that's a lot, but here are some of the main ones that apply to investing you should be on the lookout for.

Hindsight Bias

This is exactly what it sounds like. It's the tendency to overestimate your ability to predict an outcome after the fact. "I knew this (terrible thing) was going to happen all along." No, you didn't. This is exactly what people said about the financial crisis in 2008, but nobody had a clue, or it wouldn't have caused so much damage.

Hindsight is 20/20, so the next time you think, "Oh, I knew that was going to happen," test yourself. Ask what you would have done had you *really* known the outcome. Because if you *really* knew your team was going to win the Super Bowl, you would have mortgaged your house to put a half-million dollars down on the game. If you had *really* known, you would have taken advantage of your knowledge. If you *really* knew what the right stock was in advance, you wouldn't have bought any other stocks. You would have put everything you owned into that one stock you knew would be the best. But you had no idea, so you didn't do any of that.

Hindsight bias tricks us into believing we have more skill than we do, which makes us more susceptible to irrational behavior going forward.

False Patterning

This is the tendency to mistakenly perceive connections and meaning between unrelated things that do not exist. This is often the basis of superstition. If I do x, then y is more likely to occur. It's why Kansas City Chiefs quarterback Patrick Mahomes has worn the same underwear to every game of his NFL career. How much does that help? His brain may think it does, but in reality, x has nothing to do with y. The problem is that our brains don't want to accept randomness for what it is. Randomness is boring, so when events occur in clusters, we seek out an explanation in the form of a pattern.

Sometimes we look for patterns to avoid and other times we look for patterns to pursue. Ten thousand years ago, if you spotted a tiger in a nearby cave on three separate occasions, you'd probably avoid that cave. But, if you saw a herd of woolly mammoth make their way across the valley right before the onset of winter, you knew you could hunt that food source. Some patterns we avoid; others we run toward. That's a useful instinct to develop, and for a long time throughout human history patterns were essential for our survival. Today, if you see that a mutual fund has outperformed the market three years in a row, you'd be more likely to put all of your money into that fund. That is not a useful instinct.

Today, we live in a different time with different circumstances, but our reasoning hasn't adapted. We still live in survival mode, and our pattern-finding machines. The problem is that we're now finding patterns that don't exist. That's why we must rely on statistics and math. For example, look at all that goes into the process of testing a new prescription drug for the marketplace. You can't know its viability and the impact it has on real patients without statistics. The same is true with investing. We must rely on science and math because false patterns are everywhere.

Prestige Bias

This is a big one I see with investors all of the time, and it occurs when people imitate investing models they believe will confirm their status and power. Meir Statman says, "Status seeking is very normal. People will deny it for themselves, but they can see it in other people."[6] In other words, this one can creep up on you, but if you can spot it, you got it.

As humans, we don't come right out and say how wealthy and successful we are—at least most of us don't. Instead, we surround ourselves with material possessions and status symbols that say those things for us. It's one reason why people join country clubs, purchase houses in a certain area, or drive a specific type of car. It also influences how people invest their money. This is how Bernie Madoff got away with such egregious crimes for so long because investing with Bernie was considered "prestigious."

Confirmation Bias

There is a good chance you've been doing this since you first cracked open this book. It involves only being able to see information that supports your current position and beliefs. We're always looking for evidence that tells us we're right. And we all want to be right. We want to be right so badly that we will favor certain information and disregard other information to confirm those preexisting beliefs. That's why two people can witness the same event and have two completely different interpretations of what happened. Think about a murder trial with video evidence and eye-witness testimony used to determine if the defendant committed murder or an act of self-defense. People can look at the same set of facts and have two very different opinions.

We always see the world through our screens and rely on what we already believe to interpret the information we perceive. That's why confirmation bias is so toxic. You've already made up your mind, and that prevents you from considering any scientific or mathematical evidence that might threaten those beliefs.

Overconfidence Bias

This is an undeserved exaggerated certainty or confidence in one's skill or ability. There have been times when we've all been overconfident about something. It's true; we are often delusional about our ability. I'll prove it to you. Do you think you're a good driver?

Most people answer yes to that question, and if you did as well, you aren't alone. According to a AAA study,[7] 73% of US drivers consider themselves to be above-average drivers—80% of men. Do the math; more than half the population can't be better than average at anything. I do this exercise during workshops, and sometimes, as many as 90% of the people in the room claim to be above-average drivers.

People can remain overconfident even when you tell them they will probably be overconfident about something. It's just part of our DNA. So many investors are overconfident about how their portfolio will do. That bias prevents them from ever recognizing the flaws in their approach.

Recency Bias

This runs rampant throughout the industry. It's the human tendency to rely much too heavily on recent data. People put a lot of emphasis on the previous three years while disregarding 75 years of academic data. That's because the recent past is vivid in our minds. If we saw something in the news yesterday or watch an investment go down in real time, it makes a clear impression.

This led to disasters like the dot-com bubble. At the time, people thought you were crazy if you didn't own tech stocks. Why? The average investor was drawing on five years' worth of returns, during which the S&P saw 28% or greater growth. Tech stocks as represented by the NASDAQ Index were up over 45% annualized return from January 1995 through December 1999. However, that same vividness can be the result of an extreme situation. If you lost half of your money in that same dot-com bubble, that experience will remain vivid and can just as easily skew your judgment when making future investing decisions.

Here's the problem: by the time the stock is up, you've already missed the returns. It sounds paradoxical, which is why so many average investors fail to grasp this concept. Investors who chase returns don't necessarily get returns. You must purchase the stock before it goes up, not when it's already up. That's why people who put most of their money into tech stocks when they were already up likely saw significant declines in their investment as the NASDAQ lost over 75% from January 2000 through September 2002. Like the movie *Groundhog Day*, this pattern seems to repeat itself over and over for many investors.

Herding Bias

This is the phenomenon where you blindly follow what you perceive others are doing. If you're in the water at the beach, and someone yells "shark," it's a good idea to rush to shore with everyone else. Your brain will propel you to do this without any further analysis. It identifies danger, and you naturally react, so there is a reason you have this bias. We're hardwired to look for danger, which often means following the pack. It can save your life in certain situations. But it doesn't work when investing because there is no safety in numbers.

Herding bias is one of the reasons why so many people dump their wealth into Bitcoin. As mentioned, $2 trillion has been wiped off the value of cryptocurrencies from its peak in November 2021 to June 2022.[8] Whether we assume that others know something we don't or don't want to miss out on what so many people believe to be a sure thing, investors instinctively gravitate toward the behaviors of the herd.

The flawed logic is that if everyone is doing it, it must be the right thing to do. Surely everyone isn't wrong, and if everyone else is getting rich, you should, too. That makes it easy for us to rationalize behavior, no matter what the science and facts might say. It's difficult enough to account for our own biases, never mind the biases, instincts, and emotions of everyone else in the herd. Herding is great for zebras, but bad for investors.

The Necessary Lie Bias

This is a justification for imprudent behavior. Necessary lies are related to what we previously discussed with the rational brain trying to excuse and explain irrational actions made during an amygdala hijacking. We do this after overeating, gambling, drinking, and engaging in any other activities and behaviors that are not in our best interest. This is where "just one more" and "I'll stop tomorrow" come into play. We also do this with stocks. "I don't care how low it goes. I just need to wait until I get even." These are all the lies we tell ourselves to indulge our preexisting beliefs and biases. It's a way to get ourselves off the hook so we don't have to take responsibility for the consequences.

Familiarity Bias

This occurs when investors create a portfolio that is biased toward familiar assets. It's the opposite approach from creating an unbiased portfolio derived from a theoretical model or empirical data. This can also happen with people. Investors will make the mistake of going all-in with managers they know without asking many questions or doing their due diligence. They do it simply because they are familiar with those managers, so they make the mistake of trusting them completely. It commonly occurs when people invest in the company they work for. That ratchets up the risk. If something happens to the company and the stock is down 50% when you lose your job, then you lose half your money and are out of work at the same time. That can be devastating. A prudent investor will decrease the risk through diversification, and that's what we will cover in Part III.

Fear of Missing Out

You've heard about the fear of missing out, but you might not realize it's also a cognitive bias. It's similar to herding bias, but in this case, it isn't the herd influencing you; it might only be one person. If you hear about your friend making a huge amount of money investing in Bitcoin or a certain stock, it's natural to feel tempted. So, if you stick to your investing principles and don't get on board with what everyone else is doing, and the stock makes money, you can't help but feel that you missed out. Those experiences leave an impression, so the next time a similar opportunity arises, you might be more likely to give in, and there is no way to know if events will turn out in your favor. Press your luck enough times, and you're bound to lose eventually.

****

None of these biases occur in isolation. You must potentially contend with all 175 biases at the same time. They all influence each other. *"Not only is everyone investing in Apple, but it's a large company I'm familiar with, so I want to make money from that stock as well."* That type of thinking becomes difficult to combat as an investor. And don't forget that professional money managers struggle with all of the same biases, emotions, and instincts.

I knew an advisor who managed approximately $100 million. She claimed to have the discipline and chops to avoid the behavioral pitfalls. But when the market crashed in 2008–2009, instead of staying strong and rebalancing by selling fixed income to buy more equities, she sold every single one of her clients' investments and went to cash. It was her job to protect her clients and keep them disciplined, but she couldn't remain disciplined herself. She's not alone. This happens all of the time.

For an advisor to remain in business, they must keep their clients. So, when their revenue drops by 30% because the market tanked, they not only have a business to run and a mortgage to pay but they also have the added stress and anxiety of employee and client concerns. In my opinion, advisors overreact even more than the average investor because their livelihood is at risk. That's why you can't rely on most advisors to protect you from their own biases.

Psychologist and economist Daniel Kahneman, the author of *Thinking Fast and Slow*, one of the foundational books on behavioral science, admitted that he still wasn't any better at managing his own money and behavior than the typical investor. Knowledge itself is an excellent starting point but it's not enough because it can't eliminate the human bias, emotions, and instincts that lead to poor investing decisions. It's how our brain is hardwired. And if that isn't bad enough, we then find additional ways to justify our behavior through the stories we tell ourselves.

Story Is Not Science

What's so frightening about these biases is how, even once we know they exist, we can still succumb to them without even realizing it. The reason is because all of these biases come with a story we tell ourselves. Stories are great. Everybody loves a good story. And we all have a story about why we do what we do in life, but that's not science. Investing is no different.

Most of us believe we are logical when making important investing decisions. However, all of these decisions are influenced by our emotions and biases much more than science. There is a good chance you own the stocks in your portfolio because there is a good story attached to that stock. Tesla has a fascinating story, but where's the academic science that says it's a good investment? In early 2024, it was one of the poorest-performing stocks on the S&P 500.[9] The point is, when investing in individual stocks the risk is often massively higher than the index that it belongs to.

Applying "the science" requires following the scientific method. That begins with an observation or a question. You then study the evidence, conduct research, and make assertions to develop a hypothesis. You then test that hypothesis by running experiments and analyzing the data. Stories can't provide proof. At best, stories can provide anecdotal examples, but that's not the same as hard evidence. When I talk about the academics of investing, I'm talking about using the scientific method to study the brain, economics, and finance to create portfolios. Unfortunately, scientific evidence is not nearly as appealing to our brain as a good story.

We become attached to our stories and committed to the stocks that are a part of them. We believe that we're prudently investing by telling ourselves a story that we're smarter or more informed than other investors. This is an especially dangerous story because it prevents us from ever accepting that there might be a better way of learning from our mistakes. These stories can convince us that this time or this situation will be different.

Unfortunately, common sense doesn't protect us from our stories or our cognitive biases, and it's not your fault. Biases originate in our subconscious, and 95% of our brain activity occurs in the subconscious. That's why it's so difficult for people to change. You're not aware of it, and you're not aware you're not aware of it. You don't have any more control over your biases than you do in growing your fingernails. It's just happening, whether you want it to or not.

It's difficult to see your own biases, which is why it's a huge breakthrough when they become visible. That means those biases no longer have control over you. You can't do anything about a problem you don't know exists. Once you can finally see the problem, you can take the first step to becoming more responsible for your future. That's why I believe that investing is a process that helps you better discover yourself.

I promise that the most valuable benefit you will get from this book has nothing to do with money and everything to do with what you learn about how your mind works. Those are the dividends and real breakthroughs because they will influence everything you do in life. But don't assume that discovering your biases will make them disappear. Biases will always be present in your life.

That's just part of being human. However, the more familiar you are with those biases, the better you can recognize what you're doing (or not doing) to better live your purpose. That makes it harder to believe the story and buy into the lie your brain is trying to tell you.

So take a good hard look. What cognitive biases to you succumb to in your life?

Notes

1. Anil Gulati, "Understanding Neurogenesis in the Adult Human Brain," *Indian Journal of Pharmacology* 47, no. 6 (November–December 2015): 583–84. https://www.ncbi.nlm.nih.gov/pmc/articles/PMC4689008/#:~:text=Each%20neuron%20can%20make%20connections,approximately%2060%20trillion%20neuronal%20connections.
2. Catherine Caruso, "A New Field of Neuroscience Aims to Map Connections in the Brain," Harvard Medical School, January 19, 2023. https://hms.harvard.edu/news/new-field-neuroscience-aims-map-connections-brain#:~:text=In%20the%20human%20brain%2C%20some,the%20human%20brain%20to%20fathom.
3. "Obesity in the U.S.," Food, Research & Action Center, 2024. https://frac.org/obesity-health/obesity-u-s-2#:~:text=The%20latest%20data%20indicate%20that,tend%20to%20increase%20with%20age.
4. "The Power of the Subconscious," Accountancy Ireland, August 26, 2022. https://www.charteredaccountants.ie/Accountancy-Ireland/Articles2/News/Latest-News/the-power-of-the-subconscious#:~:text=The%20subconscious%20mind%20makes%20up,full%20potential%2C%20personally%20and%20professionally.
5. "Every Single Cognitive Bias," Visual Capitalist, n.d. www.visualcapitalist.com/every-single-cognitive-bias.
6. Ibid.
7. Matthew DeBord, "Americans Are Overconfident in Their Driving Skills—But They Are About to Get a Harsh Reality Check,"

Business Insider, June 25, 2018. https://www.businessinsider.com/americans-are-overconfident-in-their-driving-skills-2018-1.

8. Rob Minto, "Cryptocurrencies Lose $2 Trillion in Value Since 2021 Peak, *Newsweek,* June 13, 2022. https://www.newsweek.com/cryptocurrencies-lose-trillion-value-since-peak-1715207.

9. Melvin Backman, "Tesla Is the Worst-Performing Stock in the S&P 500 This Year," Quartz, February 7, 2024. https://qz.com/tesla-worst-stock-performer-musk-1851227426.

CHAPTER 7

What You Think You Know
Might Not Be So

"Today is an opportunity to see something new or see something in a new way."

—Donald T. Iannone

"What did you once believe that I you now know is not true?"

Sir Isaac Newton invented calculus, discovered gravity, and is considered one of the most brilliant people ever to walk the face of the earth. You would think he could use his intelligence to make a killing in the stock market. Well, so did he.

In the 1720s, Newton invested all his money in one stock—the South Sea Company. He doubled his investment in six months, sold, and got out. However, his friends remained invested, and he saw them continue to make money, so he bought back in at three times the price of his original stock purchase. That's when the bubble burst, and the stock price plummeted. By the time he sold, he had lost almost all of his money—the equivalent of $3 million today. At age 78, Sir Isaac Newton was wiped out. He managed to live the rest of his life comfortably on a government salary as master of the royal mint, and his professional legacy remained untouched. Still, he remained haunted by his stock market blunder. In a private

139

letter written three years before he died, Newton wrote, "I've lost very much by the South Sea Company, which makes my pockets empty and my mind averse from dealing in these matters."[1]

This was well before behavioral science, but it's a classic example of confirmation bias, herding bias, and fear of missing out There is a formula for gravity, yet none for human emotions. And as Newton learned the hard way, there is no safety in numbers. We work hard to concoct clever stories about why an investment won't be lost. That was the case with the South Sea Company. But beneath the surface, it was a classic pyramid scheme. Nobody ever thinks they will fall for a pyramid scheme, yet one of the smartest people to ever live did just that.

Centuries before the moon landing, the physics Newton pioneered is what eventually sent planetary probes into space and put men on the moon. He could calculate the motion of heavenly bodies and the workings of the universe but didn't understand stock market bubbles or the inner workings of the human mind. This was a man with an IQ of 190 who could predict where a planet would be in orbit one year in the future, but he couldn't predict where the market would be one day later. This isn't anything against Newton. Nobody has figured out a way to hack the market, no matter how smart they are. If anyone tells you that they can, run, do not walk, in the other direction.

Sometimes, your intelligence can work against you as an investor. It doesn't matter how successful you are; knowledge in one area does not make you an expert in another. This is an easy error to make. No matter how successful you are in your career, you would never in a million years assume you could hop into the cockpit of a 747 and fly the plane without any training. However, this is extremely common when it comes to investing. I refer to this as an attribution error. Attribution is the action of regarding something as being caused by a person or thing. In this context, it occurs when people experience success in one area, and assume that same success will carry over to investing. People read a few articles, watch some videos, and suddenly think they can engineer a portfolio that will help them consistently beat the market, sadly to find out their genius as a brain surgeon did not translate to investing success.

It's time to put your knowledge to the test. The following three exercises are a great way to determine how much you actually understand about investing and your portfolio. These have been designed to help you see how the conscious mind isn't calling the shots. If you're still experiencing cognitive dissonance or doubting any of the concepts discussed, consider this chapter the ace in the hole that will force you to test the limits of your claims.

Pick Any Three Stocks

If you had a million dollars to invest, and you had to put that money into three separate stocks, which three would you pick? Close your eyes for a moment, choose three stocks, and then write them down before you continue reading on.

When I do this exercise at workshops, most people not only pick one of these stocks that I'm about to mention but pick stocks only from this list. And this is after I taught them about cognitive biases. There are over 20,000 stocks you could choose from, and I can still list five that end up on most people's lists. What are the odds of that? Let that sink in. But first, let's see how you did. Were any of the stocks on your list Apple, Microsoft, Amazon, Google (Alphabet), or Tesla? If I was going to add a sixth, I would say Facebook (Meta). If you didn't pick any of those, congratulations on being in the minority and bucking the trend! Unfortunately, that does not mean your picks are market-beating winners.

If you did pick one or more of these stocks, don't worry. This is your subconscious mind at work. You didn't pick those stocks; you only think you picked them. You have been programmed to pick them. Better yet, those stocks picked you, not the other way around. It's a knee-jerk reaction, like the one you have when your doctor tests your reflexes by tapping your knee with the rubber reflex hammer. It all goes back to what I said at the beginning of this part: you have very little control. If any one of those familiar stocks wound up on your list, science didn't influence your decision. You picked something you were familiar with, something you knew many people were invested in, or stocks you believed performed well in the recent past. Everyone loves Apple products, so it's natural to want Apple stock. You might love what Elon Musk said on Joe Rogan, so you want Tesla stock. It's the result of every single bias we just went through—hindsight, prestige, familiarity, and herding bias. You knew about these biases—probably even recognized how they creep into your decision-making—and they still call the shots! But this instinct isn't all bad.

We are emotional beings who make emotional decisions, and that can be a very good thing. Life would be boring without emotion because we'd all act like robots. Think about your spouse or partner? You didn't need to break out a spreadsheet to figure out if you wanted to be with the person you love, and if you did you probably need to get rid of it. It's the same reason why you take care of your kids and your aging parents. Those decisions are driven by emotion, and that's part of what makes life rich and amazing. You just don't want to rely on emotion when picking your investments.

These companies may be big and well known, but in recent years they have not had the highest returns. The only reason you would ever pick stocks to begin with is because you believe you can pick the best ones. Why else would you do it? If you had that ability, which stocks would you pick? You'd pick the top five, and when I'm writing this in 2024, the actual top five stocks over the past 10 years are listed in Table 7.1, along with where Apple, Microsoft, Google, Amazon, and Tesla fall on that list.

If you had held those five popular stocks over the previous 10 years, you would have seen a 24.88% annualized return. Not bad. But if you had held the actual top five stocks of the previous 10 years in your portfolio, you would have earned over 114.75% a year. If you think that's substantial, look at the results in Table 7.2 when we look at the previous three years.

You read that correctly. The top five stocks over the previous three years saw an average 3,516.95% increase per year. How many portfolios do you think contained all five of those individual stocks? It's probably safe to say that most people have never heard of

Table 7.1   10-Year Average Return of Five Large and Popular Stocks Versus the Five Top Performers

1/1/2014 – 12/31/2023

5 Large & Popular Companies	10 Year Performance Rank	5 Top Performers	10 Year Performance Rank
Apple Inc	323	Tritent Intl Corp	1
Microsoft Corp	292	Next Meats Holdings Inc	2
Google	1,464	Eastern Communications	3
Amazon.com Inc	421	General Enterprise Ventures	4
Tesla	126	Stratos Renewables	5
Average Return of five Large and Popular Companies	24.88%	Average Return of Top five Performers	114.75%

Note: PAST PERFORMANCE IS NO GUARANTEE OF FUTURE RESULTS. This information is for educational purposes only and should not be used as investment advice. See Appendix II for additional information. The return data shown does not include the deduction of fees or expenses associated with trading, which would lower returns.

Table 7.2   Three-Year Average Return for Five Large and Popular Stocks Versus the Five Top Performers

1/1/2021 – 12/31/2023

5 Large & Popular Companies	3 Year Performance Rank	5 Top Performers	3 Year Performance Rank
Apple Inc	2,869	NEO Battery Materials Ltd	1
Microsoft Corp	2,021	Intercorp Peru Ltd	2
Google	2,400	Interfoundry Inc	3
Amazon.com Inc	6,939	Synnex Technology International Corp ADR	4
Tesla	5,518	Gold Rock Holdings Inc	5
Average Return of five Large and Popular Companies	10.01%	Average Return of Top five Performers	3,516.95%

Note: PAST PERFORMANCE IS NO GUARANTEE OF FUTURE RESULTS. This information is for educational purposes only and should not be used as investment advice. See Appendix II for additional information. The return data shown does not include the deduction of fees or expenses associated with trading, which would lower returns.

these companies. If somebody had known ahead of time that these would be the best performing stocks, that person would likely be the richest person on the planet. And if you look at this data and suddenly want to invest in NEO Battery Materials, you're missing the point because if a stock is already up, you missed out.

If you were to invest in the top 10 stocks at the end of the year, historically, the following year your return would drop below 1%. Nothing makes this point more obvious than what is shown in Figure 7.1.

Another reason why investors gravitate toward these major companies with such a large cultural footprint is that they believe these companies are too big to fail. That makes them a safe investment in the mind of many investors, but that couldn't be further from the truth. History has taught us that no company is too big to fail.

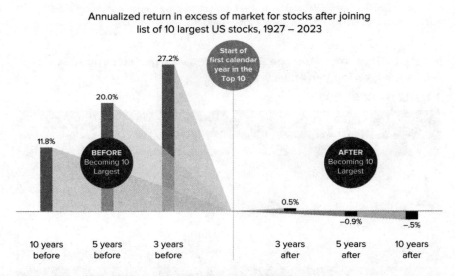

Annualized return in excess of market for stocks after joining list of 10 largest US stocks, 1927 – 2023

Figure 7.1   Strong Performance Unlikely to Continue

Note: PAST PERFORMANCE IS NOT A GUARANTEE OF FUTURE RESULTS. Data from CRSP and Compustat. Companies are sorted every January by beginning of month market capitalization to identify first time entrants into the 10 largest stocks. Market defined as Fama/French US Total Market Research Index. The Fama/French indices represent academic concepts that may be used in portfolio construction and are not available for direct investment or for use as a benchmark.

The Dutch East India Company was the first publicly traded company in the world. It had 50,000 employees. They even had their own army (10,000 private soldiers) and navy (150 merchant ships, 40 warships). If you owned that stock, it paid an 18% dividend for almost 200 years. How about that for a boom? If there was ever a company that was too big to fail, it was this one. The Dutch East India Company would be worth $7.8 trillion in today's dollars. To put that in perspective, Apple is the biggest company today, worth roughly $2.9 trillion. Despite being worth that much, the Dutch East India Company went bankrupt in 1799.

You can look at companies like Facebook (Meta), Tesla, and Microsoft and think they are too big to fail, but the truth is that no company is too big to fail. Look at Lehman Brothers, GM, Conesco, WaMu, CIT, Chrysler Worldcom, Enron, and the Pacific Gas and Electric Company. All of these companies were considered too big to fail—until they weren't.

Remember Kodak? The Eastman Kodak Company was established in 1880. It remained on the Dow Jones for 74 years. Apple, Microsoft, Tesla, Facebook (Meta)—none of those companies have been around for that long. In 1962, Kodak employed 75,000 people and accumulated more than $1 billion in sales. In 1976, Kodak was responsible for 90% of all film sales and 85% of all camera sales.

When I was a kid, we always drove down to Florida on family vacations in August when it was 95 degrees with 95% humidity. We'd take a ton of pictures, and when we got home, we'd drop them off at the pharmacy. Two weeks later, we'd go back to pick them up. Kodak then came out with 24-hour film development services from their little yellow huts. At the time, it felt like revolutionary technology. Granted, sometimes they'd make a mistake and send us home with someone else's vacation pictures, but business was booming for Kodak. People would always take pictures, so how could a company like Kodak ever go out of business?

Here's the kicker. In 1975, Kodak invented the digital camera. Employee Steven Sasson created the first portable, battery-operated, self-contained digital camera. It weighed eight pounds and wasn't practical yet, but they had the technology. Know what the CEO did?

He insisted that Kodak was a paper and chemical company—not a technology company. He felt the digital camera would cannibalize film sales, so the company never fully embraced the technology. That made room for its competitors to pass them by. In 2004, Kodak was delisted from Dow Jones Industrial Average. In 2010, it dropped out of the S&P500. In 2012, it went bankrupt.

Capitalism waits for no one. The Kodak CEO failed to see that if he didn't embrace the technology, another company would. With my company, we aim to reinvent our business model every three years. We must continually innovate because, it doesn't matter what you do, the old model will not last forever. That's why innovation and creativity are part of the American Dream. Your business can innovate, win over customers, and capture the market share, but it works both ways because your company can also go bankrupt. Kodak isn't alone. There are many other examples of companies who were unable to see or embrace change.

Blockbuster is a company that has a similar story and suffered a similar fate. In 2008, their CEO famously said, "Neither RedBox nor Netflix are even on the radar screen in terms of competition."[2] Less than two years later, Blockbuster went bankrupt because they failed to embrace technology.

I can't say it enough: no company is too big to fail. Despite seeing this play out repeatedly, this is a lesson that people never seem to learn. So, they are doomed to keep making the same mistake of investing in companies they feel are safe because of the company's size and reputation. For those of you who weren't alive when we had to develop actual pictures or were too young to ever visit a Blockbuster Video store, it's easy to say those companies were always doomed to fail. It seems obvious now—but it wasn't at the time. That's hindsight bias. When people had to wait in line to drop their pictures off at the busy yellow hut, nobody knew that one of the world's biggest and most successful companies would go bankrupt.

Jump ahead to 2018, and those who invested in Boeing felt secure because it was one of the strongest companies in the world. One of my workshop attendees had millions in Boeing. He thought the company was safe, and his future secure, so there was no need to use academic investing principles or diversification, or so he thought. Then, a failure in the autopilot system resulted in two separate crashes between 2018 and 2019 that killed 346 people. As a result, the Boeing 737 MAX was grounded worldwide between March 2019 and December 2020.[3] Then, COVID hit. The airports shut down, and the stock crashed. Over the next four years, there were a string of incidents that kept Boeing in the news (not for good reasons), including an incident in January 2024 when the door plug of a Boeing 737 MAX 9 came off midflight, creating a large hole in the side of the plane. For those who remain invested throughout all of the turmoil, the stock still hasn't recovered (see Figure 7.2).

Those of you invested in Apple are probably thinking the company is rock solid today. The problem is that technology is advancing at a much faster and unprecedented rate, so the next big product, service, or company is right around the corner. Who

Boeing Company
GROWTH OF $100K GROSS OF FEES (4/2019 – 3/2024)

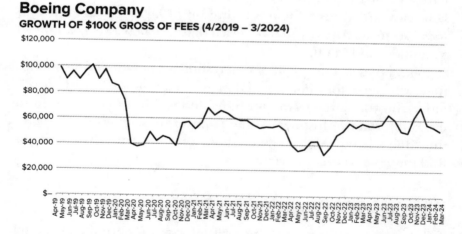

Figure 7.2   Boeing Company Stock Performance

Note: PAST PERFORMANCE IS NO GUARANTEE OF FUTURE RESULTS. This information is for educational purposes only and should not be used as investment advice.
See Appendix II for additional information. The return data shown does not include the deduction of fees or expenses associated with trading, which would lower returns.

knows what that will be, and who knows what impact that will have on the market? Add this to the long list of reasons why stock picking is so destructive.

How Much Have You Saved for Retirement?

Forget about what it takes to be considered wealthy; what do you need to be financially comfortable?

According to the 2023 Schwab Modern Wealth Survey,[4] Americans said it takes roughly $2.2 million to be wealthy, yet 65% of respondents have no formal financial plan. Of those 65% without a plan, 44% believe that don't have enough money to need a plan. That shows there is a massive gap between what Americans say they need and what they actually have.

Consider what I refer to as the *million dollar dilemma*. There is a rule of thumb in financial planning, often called the *4% rule,* where you can only take out 4% of the money you have saved each year during retirement to avoid running out of money. So, if you have a million dollars, you could take out $40,000 a year to live on. How easy would it be for you to live on $40,000 a year, even with social security? And if you want twice that, $80,000 a year, you need $2 million, and so on.

The point is that people underestimate the amount of money they will need for their golden years to be golden. How much money do you have saved? How much do you feel that you need to live comfortably? Most people can't answer those questions because they don't know. And many of those who do have answers don't find those answers comforting.

What's Your Average Return?

Talk to most investors, and they will tell you they are doing "pretty good." However, "pretty good" isn't good enough for your family's future. There is nothing scientific about pretty good. Here's the problem: most investors aren't even doing "pretty good." Most investors have no clue about how their investments are doing.

According to Dalbar Research,[5] the average person investing in equities between 1994 and 2023 (30 years) earned an average return of 8.01%. That significantly underperformed the S&P 500 at 10.15% during that same period. An additional 2% or 3% may not sound like a lot, but over time, it's a significant amount. In this example, if you invested $100,000 back in 1994, it's the difference between having $1,009,064 and $1,817,753. Given that information, what's funny is that most people still assume they are doing well. But compared to what? Not the overall market, because most investors aren't beating it and don't seem to realize it. Never once in my 40 years working in this business has a client come to me and said, "I'm in trouble; I'm underperforming the market."

The truth is that the average investor is doing poorly according to the Dalbar research. They just don't know it, and they don't know they don't know it. The better question is not whether you failed to beat the market; it's how would you even know?

To figure that out, you should determine your performance based on a time-weighted rate of return. The problem many investors make is they measure their long-term success by yearly average returns. So, they might believe they've earned 12% for the past 20 years when that isn't the case. Average rates of return can be very deceiving. Let's say you invest $100,000 and make a 100% return the first year to bring you to $200,000. In year two, you lose 50%, or $100,000. One hundred percent minus 50% is 50%. Divide that by two to get a yearly return, and it results in an average return of 25%. But how much money do you actually have? You have $100,000. You have a 25% average return, but you're right back to where you started and didn't make any money. The real return: the compound dollar, time-weighted return is zero. Many people misunderstand that, and advisors want you to misunderstand it. This is one of the reasons why investors are in the dark and are overconfident about their returns.

Mutual fund performance is another misunderstood metric that gives investors the illusion they are getting higher returns than they often are. For example, if you look up your fund on Morningstar and see that it's made 13% over 10 years, that doesn't mean everyone in that fund has made 13% on their investment. Only those who were invested for the entire 10 years earned that return. If you bought in when the fund was high, and then it went down, you could have actually lost money. Very few investors are there for the ride up, but most are there for the ride down. You don't earn the returns if you get in late. Our brain doesn't always take that into consideration. We just see the 13% and assume that's what we're getting because we're invested in that fund. The shorthand math we do in our heads to calculate our return is often severely flawed.

Another way we trick ourselves into thinking our return is higher is through our tendency to only keep the stocks that perform well and sell the ones that don't. Let's say you start off with 50 stocks in your portfolio. At the end of year one, half perform well, and half don't, so you sell the bad stocks. You repeat this pattern over 10 years, and at that point, your statement might show that 10 of the 20 stocks you own are crushing it. But you forgot about all of the bad stocks you purchased over the past decade that destroyed your overall return. This is called *survivorship bias*. By only remembering the winners and forgetting about the losers you become overconfident and think that you're good at stock picking when your track record is terrible.

Mutual fund companies will do something similar. They'll have dozens of funds, and then close the ones that don't perform well. They even go so far as to remove them from the record, so you won't know they ever existed. They understand survivorship bias and are happy to use it against investors. They do this on purpose, and that's evil, but most investors don't understand anything about survivorship bias. They just keep the winners and sell the losers without factoring those losses into their overall portfolio performance.

Is Your Portfolio Toxic?

One major reason why people don't know what they think they do is because they listen to the wrong people. They see billionaires who made their fortunes in tech, real estate, hedge funds, or oil and gas and assume that if they can mimic their strategies, they can get rich, too. It's a lie to say that if you want to be a billionaire, invest like a billionaire. The industry perpetuates this belief, and it's arguably one of the biggest misconceptions of investing— people have been watching too much *Shark Tank*. Many of these wealthy gurus imply that they will give away secrets during their workshops, master classes, and, yes, their books. The only secret is there is no secret.

In his 2024 book *The Holy Grail of Investing*, Tony Robbins promises to reveal the secrets and strategies that will enable the average person access to investments that were once only accessible to the investing elite. He says they can earn returns upward of 20% to 30% per year by investing in vehicles such as hedge funds, venture capital, professional sports teams, private equity, and private lending. Sounds great, right?

Spoiler alert: it's just another illusion. Those gurus pitching these get-rich-quick schemes make it seem like you will get huge returns, but they downplay the risks, which are enormous. These are gunslinger investments that look sexy and feed your ego, but the average investor has no business being involved in these. What these gurus don't tell you is that there are thousands of people who went bankrupt making these same types of risky investments. There is a reason why many are unregulated and not offered through fiduciaries. Some are just repackaged limited partnerships that we sold back in the 1980s. And very little of it worked then. All it does is tie up your money, and the fees in these investments go as high as 2% per year, and 20% of any gains—what's known as 2 and 20. No investor should shoulder that burden. If you have billions of dollars, you can afford to roll the dice and make risky bets. Maybe you get lucky, but the average investor doesn't have the luxury to throw money away on get-rich-quick schemes.

The single best way you can protect yourself is by being able to identify toxic investments peddled by gurus like Tony so you can immediately remove them from your portfolio, or better yet, never be duped into purchasing them to begin with. Cryptocurrency, gold, commodities, most ETFs (BlackRock offers more than 400), NFTs (nonfungible tokens), and ESG investing (environmental, social, governance) investing, all fall into this category. Buyer beware. These vehicles are speculating and gambling on steroids because there is a rush that comes with these investments that makes you want to invest more.

As mentioned, "there ain't no free lunch," and there certainly is no "holy grail of investing." That's hyperbole. It's dangerous and it's inappropriate to talk like that. No respectable investing professional should ever use that language. It creates false expectations. There is a reason that there is no hyperbole in academic investing science. You can't create wealth without working hard. It requires knowledge, determination, courage, and time. There is no alternative to prudent investing, and there is no low-risk, high-return investment. This is a difficult lesson for investors to learn. People prefer an easy lie over the hard truth. The sexier the strategy, the more seductive it becomes. Remember, if someone could really help you make a 20% return, they would borrow money at 5%, and pocket the remaining 15%. And they would never tell you!

****

These exercises and questions are always eye-opening for investors because it forces them to confront what they don't know they don't know. That's what holds so many people back and prevents them from changing their beliefs. That's why I want to end this chapter by revisiting the question first posed at the beginning: *What did I once believe that I now know is not true?*

Before we move on to Part III, let's take a closer look at what you've done wrong in the past. More important, let's determine why it's wrong. This can help you sever ties with your old way of thinking for good so you won't make the same mistakes in the future. It will also help make you more receptive to the new world of investing you will soon discover.

Can you identify at least one thing that you once believed, that you now know is not true?

Notes

1. Andrew Odlyzko, "Isaac Newton and the Perils of the Financial South Sea," *Physics Today* 73, no. 7 (2020): 30–36. https://pubs.aip.org/physicstoday/article/73/7/30/800801/Isaac-Newton-and-the-perils-of-the-financial-South.

2. Rick Munarriz, "Blockbuster CEO Has Answers," The Motley Fool, April 5, 2017. https://www.fool.com/investing/general/2008/12/10/blockbuster-ceo-has-answers.aspx.

3. Bill George, "Why Boeing's Problems with the 737 MAX Began More Than 25 Years Ago," Harvard Business School, January 24, 2024. https://hbswk.hbs.edu/item/why-boeings-problems-with-737-max-began-more-than-25-years-ago#:~:text=The%20flaws%20in%20the%20software,the%20deaths%20of%20346%20people.

4. Charles Schwab Modern Wealth Survey 2023, https://content.schwab.com/web/retail/public/about-schwab/schwab_modern_wealth_survey_2023_findings.pdf.

5. "Quantitative Analysis of Investor Behavior," Dalbar, 2023. https://www.dalbar.com/catalog/product/5.

The Five Discoveries

"The intellect has little to do on the road to discovery. There comes a leap in consciousness, call it intuition or what you will, and the solution comes to you, and you don't know how or why. All great discoveries are made in this way."

—*Albert Einstein*

"What are the root causes of your investing mistakes?

Procter & Gamble headquarters is close to where I grew up in Cincinnati. In the 1980s and 1990s, P&G was the gold standard, like Apple and Google are today. My former father-in-law worked at P&G for most of his life, so he knew many of the top executives, including the CEO. When he was six months away from retirement, he had $5 million in his pension plan. The way the P&G plan worked at the time was that you could build a mix of P&G stock or fixed income, but that was it. You only had those two options.

My former father-in-law served on my company's board. He had access to the science, he had been to my workshops, and with only six months left until retirement, he still opted to keep all the money in his pension plan invested in P&G stock. That's why I told him, "Having one stock is insane, no matter how well you're doing or how much you think you know about the company."

"Come on, Mark," he said. "I know all of the academic stuff, but I also know P&G. We're global. We're diversified. We have Tide. We have Pringles. People aren't going to stop buying Pringles, are they? I helped design Pringles. I can tell you how to make a Pringle. Everyone needs our products. We make consumer goods. People will not stop buying them."

Well, in just six months, he lost 50% of everything he amassed over his 40-year career in his pension. And it happened in the blink of an eye. So, he retired with $2.5 million when he could have retired with $5 million. Blame it on familiarity bias, confirmation bias, hindsight bias, herding bias, emotional status, or the fact he didn't want to look bad in front of his CEO by selling his P&G stock. No matter the reason, that money was gone, and he wasn't at a point where he could earn it back.

He is not alone. I hear different versions of this story all the time. Long-time employees feel trapped by their company's stock options. This is a great example of stock picking when people don't realize it's stock picking. And it's one of the worst kinds of stock picking because you're playing Russian roulette with your retirement and future. These are people with knowledge of the company's inner workings, so they have faith in the stock, but we don't know nearly as much as we believe we do. Things change, and we have almost no control over those outside forces that can tank a company and its stock. That is what makes this type of investing so dangerous because one decision can wipe you out.

Investing is different from many other things in life. Take dieting, for example. You can spend three months eating a healthy diet and exercising daily. You can lose fat, gain muscle, and completely transform your body so you look and feel better, only to splurge one night and eat a ton of junk food and candy. My weakness is butter cake with a huge scoop of ice cream and a little drizzle of raspberry. When I order this at Mastro's, the waiter brags that each piece is made with an entire stick of butter. That will not help my health and fitness goals, but at that moment, I want that cake, and nothing will convince me not to eat it, so I'm okay with cheating on my diet. We all have our moments of weakness, and

I am no exception. But just because you slip up one day on your diet doesn't mean you'll throw away three months of progress. You might wake up the next morning and slip into a shame spiral, but you will get over it and can put yourself right back on track. That one piece of butter cake isn't going to give you high cholesterol, raise your blood pressure, or cause a heart attack.

Investing doesn't work the same way because one mistake can destroy you. That's all it takes. And at the time, it never feels like a mistake. It might feel like a good idea or what everyone else is doing at the time. I've watched it happen. I see people make mistakes they believe to the core of their being are the right thing to do and still lose more money than they ever thought was possible.

During the tech boom in the late 1990s, everybody wanted all US stocks. All everyone talked about was dumping money in tech stocks. There was no talk of diversification. The story about getting rich quickly kept getting more enticing because it seemed like such a sure thing. Do you know what happened next? The dot.com bubble peaked in March 2000, and the Nasdaq lost 75% between 2000 and 2002. If you were getting ready to retire and had $3 million invested in tech stocks because everyone said it was a sure thing, you likely destroyed 35 years of hard work. You may never come back from something like that. How many dreams would be saved if those people had 10 times their wealth today? I still hear stories from people who lost a fortune in the tech bubble. They still feel embarrassed, and it was all the result of cognitive bias. Human beings don't look at the 70 years of data required to create empirically tested research. They look at what happened in recent history. Typically, 3 to 5 years—10 at the most. They see recent success and want to capitalize, not realizing that one mistake can destroy everything. And most of these mistakes are made when our emotions influence us.

People like to think personal performance in life is complicated, but it's not. There are two components: what you eliminate and what you focus on instead. This chapter sums up what you need to eliminate. It's about blowing up the old way of doing things before transitioning to the new way. Much of investing success involves getting out of your own way to avoid the most common pitfalls so you don't let the bullies and your bias sabotage your returns. If you can do only that, you are already better off than most investors.

It begins by analyzing your previous behavior to better understand where you went wrong and why. What follows are the five discoveries I made 35 years ago that changed my life. Once I made these discoveries, I never looked back. That's been one of the hardest things I've ever done, and people ask how I have managed to stay disciplined when so many others in my industry failed. I believe it all goes back to that state change I experienced when I was 27 after discovering I was speculating and gambling with my clients' money. Something inside me flipped, and I knew that I needed to protect my clients from those bullies, no matter what. I also knew that I needed the right coaches to help me stick to the science because if I wasn't careful, I could easily fall victim to my own biases. Since then, I haven't been tempted to return to the old destructive way of doing things. However, the fact that I made these discoveries for myself isn't important. You must discover this and believe it yourself for it to make any difference.

This is your wake-up call. I want you to break out a pen and paper or open your laptop to keep track of those previous behaviors and mistakes when you think of examples. If you've been investing for 20, 30, or 40 years, you've experienced pitfalls and breakdowns. You've had experiences that have led to less-than-satisfactory results. This process can be humbling, but it's essential if you want to change. Otherwise, you're doomed to make the same mistakes of the past. This can help sever your attachment to the comfortable and possibly addictive way of speculating and gambling with your money.

Discovery One

Stock picking, in all its forms, is destructive behavior.

Stock picking is destructive because it forces you to take too much risk to get too little return. That puts your American Dream in danger and prevents you from ever having peace of mind.

Let's first define our terms, starting with peace of mind, because it's often misunderstood. Peace of mind is not the result of portfolio construction. Too many investors falsely believe that if they can just get the perfect mix of assets that give them high returns with low risk, they will have peace of mind, but that's not true. Peace of mind is a state of mind. Specifically, it's an understanding that you are taking the correct actions, regardless of circumstances or outcomes. You can't control the economy. You can make prudent decisions for you and your family. Knowing what you are doing and why at all times, whether your portfolio is up 20% or down 20%, is what creates peace of mind. And stock picking is antithetical to peace of mind.

Stock picking is the most common example of speculating and gambling, and it's the hardest discovery for many investors to make. However, if you've been paying attention, you should already be on board. Don't be duped by the illusion created by the bullies or get sucked into viewing investing through the friend filter. Stick to the facts. Evidence comes in the form of the consistent failed predictions by the prognosticators and gurus. Evidence comes in the form of no company being too big to fail. Evidence comes in the form of successful managers not being able to consistently repeat their performance. Evidence comes in the form of monkeys being able to predict the market better than most professionals by throwing darts. Most significantly, evidence comes in the form of your own failure to beat the market or get your desired rate of return.

If you currently hold individual stocks in your portfolio, you are most likely engaged in stock picking. But don't only consider your current portfolio. Consider your entire investing history and look for examples where you invested and lost money because of stock picking. What was the company? How much did you invest? How much did you lose? This can be painful. You need to fight the human urge to forget or minimize the impact of this experience.

In one workshop, a couple admitted to losing money through stock picking. When I asked for details, the husband said he had lost $25,000. Suddenly, the wife started crying. At first, I didn't understand what was happening, but eventually, she admitted that they lost $250,000 that they had been saving to put their children through school. That's a very big difference. Fight the urge to minimize because that's a way your brain tries to excuse or justify the mistake. And when you do that, you can't learn and grow.

Given the abundance of evidence presented so far, it may seem obvious to some that stock picking is destructive. Still, this is a difficult hurdle for many to overcome. It's addictive and sexy. And there is no better feeling than when you get it right, so some investors don't want to give it up. For others, it's been ingrained in their heads from such a young age that they struggle to change their behavior.

When my oldest daughter, Mallory, was in third grade, she came home from school one day excited, and told me, "Dad! We're playing the stock market game in school. We have $10,000 in pretend money to pick our stocks, and whoever makes the most money wins an award. Do you have any advice on what stocks to pick?"

I couldn't believe what I was hearing. This stuff is embedded deep in our culture. Our children are being brainwashed as early as the third grade to think that it's okay to gamble with their money and their future when there is not a stitch of evidence anywhere proving this is a good idea. They might as well have blackjack day at school. Both are forms of gambling and can be equally destructive.

Destructive means "causing great or irreparable harm or damage," and stock picking does that by increasing risk. Picking individual stocks is three times as risky as owning the entire asset category. It has the potential to destroy your long-term returns, and increases the chances of catastrophic outcomes, think Boeing, P&G, or Kodak-type losses.

When your investing strategy is based on a forecast, it will also prevent you from ever experiencing peace of mind because you will never stop wondering and worrying about what will happen next. This can create fear, apprehension, and mental obsession. This is true even when your predictions pan out and the stock you pick makes money. Once you see that it worked once, you're more likely to do it again. That one right prediction becomes a gateway drug that leaves you continually trying to replicate that same high. You are much more likely to experience a catastrophic loss than you are to hit the jackpot again, but you can't recognize that because all of the biases have kicked in. It's just as if you went to Vegas to play roulette. You could bet $100,000 on 36 red, which pays out 35 to 1. If the number hits, you win $3.5 million. That's incredibly lucky, and there would be no skill involved, but the rush you experienced from that lucky guess, will keep you coming back again and again.

Before you move on to the next discovery, I want you to write down these words and say them out loud: *"I have discovered for myself that stock picking in all its forms is destructive behavior."*

Discovery Two

Market timing, in all its forms, is destructive behavior.

One reason market timing doesn't work is that the market movements are so fast and furious that you can miss out on major returns if you're not invested for one particularly good day.

If you invested $1,000 in the S&P 500 in 1970, and didn't touch it for 50 years, in August 2019, that investment would be worth $138,908.[1] However, if you missed the single best day of returns during that time, you would miss out on $14,000. That's 10% of your money because you weren't in the market one day. Miss the best 25 days over 50 years, and your total investment would only be worth $32,763. You'd have $100,000 less just because you weren't invested during those 25 days. This isn't an anomaly. It is par for the course.

In 2023, the S&P was up 14% by November, with only 11 more days when the market had a positive return rather than a negative return. Most gains over those 10 months in 2023 can be attributed to eight individual days.[2] And nobody knows in advance when those profitable days will occur, which makes timing the market impossible.

Still, many believe they can get in and out of the market at the right time. When living during a period or event when the market sees expected gains or losses, people are urged to either get in or out to capitalize on the current conditions. COVID, Brexit, and election cycles all fall into this category, but the problem is that you don't know anything that someone else doesn't know. In other words, all that knowable and predictable information is already baked into the price, so only random and unpredictable events will change the price going forward. It's all market timing. I keep repeating this because it's one of the most important lessons, if not the most important lesson, in the book.

Market timing can manifest in several ways, but I will focus on two: a commodity breakdown and a market panic breakdown. As you read this section and I provide examples, consider times when you engaged in market timing that produced unfavorable results. Add these to your list and do so without judgment. Right now, we're just taking inventory.

Commodity Breakdown

Commodities have zero expected return because, except for gold and silver, they are perishable. They are used by farmers to lock in the current price to reduce their risk. You aren't buying the actual commodity; you're buying a future or an option on the commodity. They will be volatile in the short term but will eventually mature or expire. There is no potential long-term growth, which makes them a terrible investment, but that doesn't stop gurus from pushing them.

Following the crash from October 2007 to February 2009, when the S&P 500 total return was down 50%, commodities as measured by Bloomberg Commodity Total Return Index were up 57% through February 2011, so many people thought that was the solution. "Get out of stocks and get into commodities." That was a popular opinion because it was supposed to be a hedge against inflation. One of those people was Jim Rogers who was considered the commodities king on Wall Street, famous for pushing gold as a hedge against inflation. In 2010, he predicted that gold could cross $2,000 an ounce in the next 5 to 10 years. He also predicted sugar going up as well.[3] For a little while, Jim was right. The commodities index went up but peaked in February 2011, so if you had invested $100,000 when he told you to, it would have bottomed out at $46,000 over the next nine years. Oops! So much for his market timing. This is just another reason why commodities are on that list of toxic investments to avoid.

Market Panic Breakdown

This is probably the most likely breakdown people will experience regarding market timing. In 2008, our buddy Jim Cramer was on *Today* telling people to take their money out of the stock market, even if it meant taking a big loss because he—and I quote—"doesn't want people to get hurt."[4] So sincere (note the sarcasm). He's predicting the market will continue to go down.

About this same time, Howard Davidowitz went on *Closing Bell* to say there was permanent damage to our financial system and that America would never be the same. He predicted there was a 50% chance we would go into a full-scale depression that wouldn't ever be fixed.[5] In 2010, guru Tony Robbins gave similar advice to consider getting out of the stock market, while simultaneously claiming *not* to be a market timer or a financial planner. Oh, the sweet irony.

How did this advice pan out? Figure 8.1 shows the S&P 500 growth from October 2007 through December 2023, and pinpoints when each guru made these statements.

None of these gurus saw the steady increase that followed coming. If you had taken their advice, you would have missed out big. It doesn't sound like market timing when the so-called experts say it; they simply make it seem logical. Nobody knows if the market will go up or down in the short term. What we do know is that in the long run it has always gone up.

I believe in the power of free markets, and when the market is down, I believe entrepreneurs will help improve the economy. That's not a market timing prediction. That's an understanding that the market's long-term trajectory has historically risen. You just need to be in it for enough time to capitalize on those boom days nobody sees coming.

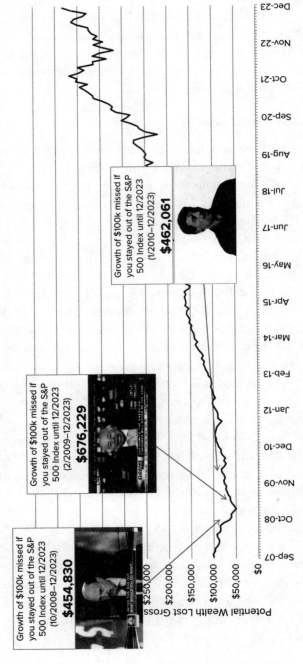

CHART: S&P 500 GROWTH OF $100K (10/2007 – 12/2023) GROSS OF FEES

Growth of $100k missed if you stayed out of the S&P 500 Index until 12/2023 (10/2008–12/2023)

$454,830

Growth of $100k missed if you stayed out of the S&P 500 Index until 12/2023 (2/2009–12/2023)

$676,229

Growth of $100k missed if you stayed out of the S&P 500 Index until 12/2023 (1/2010–12/2023)

$462,061

Potential Wealth Lost Gross

$250,000
$200,000
$150,000
$100,000
$50,000
$0

Sep-07 · Oct-08 · Nov-09 · Dec-10 · Jan-12 · Feb-13 · Mar-14 · Apr-15 · May-16 · Jun-17 · Jul-18 · Aug-19 · Sep-20 · Oct-21 · Nov-22 · Dec-23

Figure 8.1   Potential Wealth Lost

Note: See Appendix II for additional information.

164

When I run workshops, I make this point not by trying to predict the future but by proving how difficult it is to predict the past. Take a look at Figure 8.2. This shows the performance of US large stocks, US small stocks, T-bills, and international large stocks between 1970 and 1980. I reveal it one year at a time and have people predict the highest-performing asset category the following year. You can play along by using a piece of paper to cover up the results and reveal the answers year-by-year to see how well you do at predicting the future.

What's revealing about this chart is that not once between 1970 and 1980 were US large stocks the highest-performing asset category, yet that is where many investors have most of their money. And out of thousands of people who have attended my workshops over the years, I've never had a single person get every answer correct and make it all the way to the end. Look at the size of some of these returns. If somebody were able to market time and predict which asset category would have the largest return year after year, they would make a fortune.

If it's this difficult to predict the past, how can you expect to predict the future? This is discovery number two. Write this down

	YEAR	Large US	Small US	T-Bills	Large Intl
	1970	4.03%	−12.21%	**6.52%**	−10.51%
	1971	14.32%	19.48%	4.39%	**31.21%**
	1972	18.98%	3.26%	3.84%	**37.60%**
	1973	−14.67%	−36.47%	**6.93%**	−14.17%
Choose the	1974	−26.46%	−26.13%	**8.00%**	−22.15%
Highest	1975	37.21%	**63.92%**	5.80%	37.10%
Performing	1976	23.85%	**51.47%**	5.08%	3.74%
Asset Category	1977	−7.18%	18.38%	5.12%	**19.42%**
for the Year	1978	6.57%	18.33%	7.18%	**34.30%**
	1979	18.42%	**45.59%**	10.38%	6.18%
	1980	32.41%	**33.46%**	11.24%	24.43%

Figure 8.2   Playing the Game

Note: See Appendix II for additional information.

and repeat it out loud: *"I have discovered, for myself, that market timing, in all of its forms, is destructive behavior."*

Discovery Three

Track record investing, in all its forms, is destructive behavior.

By this point, you know the drill. As I run down this list of examples, identify a time in your life when track record investing produced less than favorable results. Look closely at those results and consider the negative outcomes.

Mutual Funds

One of the most common examples of track record investing occurs when you purchase a mutual fund with a strong performance history. Picture this scenario. You look at Morningstar and pick out a five-star fund or an advisor does it for you. It seems like a no-brainer, but then two, three, or four years down the road, the fund doesn't come close to performing the way it did before you purchased it. In many cases, it's not doing that well at all. You have not even beat the market. Or worse, you're losing money.

It's easy to be seduced by a manager. It's also easy to be seduced by an asset category. How often have you heard someone mention adding gold and precious metal mutual funds to their portfolio? Whenever someone asks me about adding gold, I have to remind them that I'm not aware of a single academic study that shows gold is a good investment. But it's what the gurus are trying to sell on TV, and it will continue. If you turn on cable news any time there is uncertainty in the market, which there always is, you will hear things like, "There is no ceiling to the price of gold." "If you don't own gold now, you'd better by the end of the year because that's when the market is expected to go down."

Gold may be great for jewelry, but it's bad for your portfolio. Gold has a very high standard deviation (or high volatility) and, historically, a very low return—only 4.9% based off of gold spot price data. That's it. That's all you need to know. The investors who bought into the idea that there is no ceiling to gold risked losing a lot of money when it went down back in 2011. It did come back up, but how many people had the patience to hold onto that investment for 10 more years so they could break even?

It doesn't matter what the asset category, chasing trends, and investing in what is hot now is a destructive way to invest. It's been happening for as long as there has been a market. But we can gain some much-needed perspective by studying the market over long periods, not relying on random market movements.

Investing in Only US Stocks

When I analyze portfolios, I see many people only invest in stocks in the United States and do not diversify internationally. Even John Bogle, the founder of Vanguard and creator of the index fund, said, "I wouldn't invest outside the U.S. If someone wants to invest 20% or less of their portfolio outside the U.S., that's fine. I wouldn't do it. But if you want to, that's fine."[6] That's Bogle's advice. Let's see how that would pan out if you followed it.

If you had invested the S&P 500 index fund back in 2000, by 2023 you could have made 410%. Not bad. If you invested in large value international stocks, they did less at 315%, but that's not the only international category you can invest in. If you were in the emerging markets index, you could have seen a 390% return. International small stocks saw a 542% return. Emerging markets small stocks saw a 648% return. Emerging markets value a 760% return. So, sorry, Jack. I have great respect for you, but that's not good advice.

It's all track record investing, whether you're doing it, a guru is doing it, your advisor is doing it, or even Jack Bogle is doing it. By looking at what managers and asset categories did well in the past and then skewing your portfolio in that direction, it's track record investing.

Active Management Breakdown

Investors love to look at which mutual funds did well in the past and then go all-in with those managers to invest their money in the future. I could do many different iterations of this, but I will give you one example. If you look at the top 30 US equity mutual funds from 2003 to 2012, they made an average of 18.01% per year. And that's a pretty good return because the S&P only made 8.84% a year during that period. But if you look at the average annual return of all US equity mutual funds during that same period, the returns are only 7.67%. Look at it that way, and those top managers appear to be geniuses. Shouldn't you give your money to these top mutual fund managers to invest because they perform so much better than the market? It seems obvious, but that's a terrible idea.

Did those top 30 managers who made 18.01% repeat their performance between 2013 and 2022? Nope. They only made an average of 10.21%, while the average annual return for all US equity mutual funds was only 9.65%. That sounds pretty good until you look at the S&P, which made 13.73% during that period. Those managers didn't even beat the S&P. Ultimately, the managers who are the supposed "best performers" today will most likely underperform their relative category or index moving forward. Why? Fees, expenses, trading costs, and taxes.

When I make this point during workshops, this is usually when someone raises their hand and says, "Yeah, but Mark, if I average 18% and 10%, it's still higher than the average of 8 and 13." Yes, that's correct, but there is one huge problem. Going into 2003 you had no idea which managers would be in the top 30. Unless you have Doc Brown's time machine from *Back to the Future,* you can't predict that with any certainty, so there is no guarantee you would have invested in those funds. There would have been a different list of top 30 managers, and that is exactly the point. You will never know the top performers in advance.

Ingrain this in your brain: a manager's ability to pick stocks in the past has little to no correlation with their ability to do so consistently in the future. In my opinion, it has zero correlation. The same thing applies to you if you're the one picking the stocks. Your past performance has no impact on your future performance. It's all luck, and luck is not reliably repeatable. Think back to the coin flip exercise. You will get different results if you do that same exercise again and flip a coin 10 times in a row. Out of those 100,000 people in that theoretical stadium, a completely different group will be left standing the next time. Why? Because markets are efficient and all knowable information is already baked into the pricing, so track record investing delivers returns significantly less than the market, creating massive risk to your portfolio and American Dream. Once again, this is speculating and gambling posing as investing.

Don't forget that many advisors are just as delusional and ill-informed as their clients. I know because I've trained thousands. They might be able to sell, but they are just as susceptible to cognitive bias, and many believe that they can pick the best stocks and managers in advance. Their discipline is no better than the investors. So, when Bitcoin goes up, they try to sell that. When it's gold ETFs, they push that. When the market drops, it's time to tell everyone to get out of the market.

The advisors are supposed to know better, but far too many don't. I watched many advisors steer their clients off a cliff. They just don't have the discipline required to stick to the academic principles because they, too, are human. They get scared and let their emotions influence their decisions. All the sound theories are useless if you don't have the discipline and patience to apply them. That's another reason why tips and tricks don't work and why spending time understanding the human side of investing is so important. Human behavior can sabotage the most fool-proof plan. As Mike Tyson famously said, "Everyone has a plan until you get punched in the mouth." That applies to investing as well.

Although many advisors are ignorant, as I once was, many others are also in on the con. When you can make money by scaring and manipulating people, why seek an alternative? It's difficult to convince someone to believe something when they are paid to not understand it. Those advisors who learn from their mistakes and change their approach accordingly are few and far between. It's much easier for them to delude themselves into thinking they know which stocks and asset categories will outperform the market in the future. That's why the only thing more absurd than gambling and speculating with your money is paying someone else to do it for you.

Write this down and repeat it out loud: *"I have discovered, for myself, that track record investing, in all of its forms, is destructive behavior."*

Discovery Four

I've made these mistakes in the past, and left to my own devices, I am likely to make them again in the future.

This discovery gets a little more personal. I don't want to assume, but in my experience, unless you're young and have never invested a nickel, you've likely done one or more of the above three things in the past. You should have a list of examples of stock picking, market timing, and track record investing gone wrong.

Let's take this exercise to the next level so this discovery can sink in. Go through the list of mistakes you compiled and see if you can identify which cognitive bias resulted in each decision. Rarely is there only one bias involved in a decision. Human beings are much more complicated than that. Nobody likes doing this. It's not fun to look back on missed opportunities and see how you were wrong in the past. However, the closer you look, the more you learn, and the less likely you are to repeat the same mistakes in the future. As the strategic coach Dan Sullivan said, "All growth starts with the truth." If you can't look at the truth, you will never be able to overcome your old screens or choose new ones that better serve you. It's like hitting C-12 on the jukebox to listen to the same song over and over because you will get the same results.

For those of you who are married, try putting this discovery into practice with your spouse. Look closely at your own mistakes and learn how you can be better. At the very least, I figured out that simply being able to say, "You're right, honey. I was wrong," instead of becoming defensive and argumentative has allowed me to stay married a lot longer. That's a fact. The ability to say, "I am wrong" gives you power to grow. Simply considering that you might be wrong makes it possible to learn what you don't know you don't know. It's when you think you know everything that you become locked into your beliefs and can't grow at all.

This discovery is not only about recognizing the mistakes of the past—it's about understanding how easy it is to make these mistakes again because these behaviors are so ingrained in human nature. It will feel so natural, so you must work to change that behavior. That requires coming to terms with your human vulnerabilities. I firmly believe the world would be a much better place if more people could actually look at their own failures and learn from them. That's a skill that's lacking in our society. There is no better example of this today than politicians. When was the last time you heard a politician say they were wrong about anything? We can recognize that insincerity in others, yet it's so hard to see it in ourselves. It won't always be easy, but that is what's required to truly move on.

Write this down and say it out loud: *"I have engaged in one or more of these activities in the past, and left to my own devices, I am likely to succumb to my biases and human behavior and make them again."*

If you are young, or just starting out investing, understand that you don't need to have made these mistakes to learn from them. There are two types of people in this world. The first type can only learn from their negative experience, and most fall into that category. The second type can learn from other people's negative experiences to prevent those destructive habits and beliefs from ever taking root. You can work to be the latter.

Discovery Five

It's abjectly absurd to speculate and gamble with my money.

If you want to gamble, do it for fun. I have no moral objection to gambling as a pastime or hobby, but I'm not talking about a hobby. If you continue to speculate and gamble with your investments, the consequences are much higher than simply missing out on the chance of high returns. It's costing you the freedom, fulfillment, and love that comes with an extraordinary life. It's jeopardizing your family's future because you can't speculate and gamble and experience the American Dream. Those two actions are diametrically opposed unless your purpose is gambling and ego. I'm not passing judgment, but that's not what drives most of the people I coach.

Knowing that and having made the previous four discoveries means that you understand it's abjectly absurd to continue speculating and gambling with your money. *Abject* means experienced or present to the maximum degree—it's as strong as you can get. *Absurd* means that it's utterly or obviously senseless, illogical, or untrue. It's contrary to all reason or common sense. It's laughably foolish or false. And don't be afraid to laugh at yourself. "Don't take yourself too seriously." That's what my father said all of the time, and he was right. Being able to laugh at yourself at times is such a liberating feeling. It can also help you accept the truth.

The difference between understanding this idea conceptually and discovering it for yourself comes down to your ability to see how it's manifested in your actual life. To do that, I want to conduct a thought experiment, which is a mental assessment of the implications of a hypothesis. Thought experiments are devices of the imagination used to investigate the nature of things. They are used for diverse reasons in a variety of areas including economics, history, mathematics, philosophy, and the sciences, especially physics. I love thought experiments because they force you to think clearly about the limits of your claims, and there is no more destructive claim than the idea that you, or someone else, knows what stocks will perform well in the future. When you try to predict the future, you are essentially playing God. That is the height of arrogance and delusion.

To make this point during my workshops, I ask a volunteer up on stage who isn't yet 100% on board with the five discoveries. They "want to want" to give up speculating and gambling with their money, but they just aren't there yet. Something is nagging at them or holding them back from fully getting it. This is someone who is confident in a particular stock they own. Following is a combination of multiple conversations I have had with investors over the years struggling to give up stock picking, market timing, and track record investing. It is an amalgamation of the struggles and cognitive dissonance they face.

Before you read this dialogue, I want you to pick a stock that you think you're comfortable with and play along by asking yourself these same questions. It might be the stock for the company you have worked for your entire life. It might be one of the stocks for those big companies you picked for the previous list. It might be the stock that everyone is buying, and all of the gurus on television are telling you to dump your money into it because the price will only go up. Pick a stock that you sense is "the one."

MARK	I want to thank Lauren for participating. I know this exercise will make a big difference for everyone here. I appreciate that she's brave enough to get on the field and do this in front of the entire group. It takes a big person to help the class in this way and commit to breaking the no-talk rule. I want to make sure that everyone acknowledges Lauren and gives her a round of applause.
LAUREN	Thanks!
MARK	And you have nothing to worry about. You're safe up here, and we're going to have a good time.
LAUREN	We'll see about that. I'm a little nervous.
MARK	So, what is it you're grappling with?
LAUREN	I have one stock left in my portfolio, and it's Tesla. I like Elon Musk and his quest to colonize Mars, and all of the other things he's trying to do.
MARK	That's great! I think it's important to acknowledge that you're in love with a story. And if that story had a title, it would be "Elon Musk Is Great, I Believe in Him!"
LAUREN	Yes. I believe it's going to end well, so I want to be a part of that story.
MARK	Does any part of you fear that owning this stock could be dangerous, or that it's speculating and gambling?
LAUREN	Yeah, I get that I'm gambling with my money and all that, but I do like the stock.

MARK   Then let's get you flat. First off, I agree. I think Elon is
            doing some really cool stuff. But do you know the cur-
            rent price of Tesla stock?

LAUREN   It had gotten up to 900, but right now it's around 730.

MARK   Right. So, how did it get there? Said another way, what
            determined the price?

LAUREN   People buying and selling it. The market, I suppose.

MARK   Right! Good! And what invisible mechanism is at work
            behind the scenes?

LAUREN   That's a great question, but I don't know the answer.

MARK   Let's make sure we're on the same page. We can agree
            that there are people who want to buy this stock and
            people who want to sell it. Also, people who didn't
            choose to own it all. Then there are people who would
            sell it short. You've got all of this activity going on.

LAUREN   Correct.

MARK   Adam Smith called that *the invisible hand,* but it's also
            referred to as *supply and demand.* Combine that with eve-
            rything that's theoretically knowable about that stock,
            and that determines the prices. Correct?

LAUREN   Agreed.

MARK   Now let me ask you some questions about the stock to
            see how much you really know, starting with the CEO.
            You know who the CEO is, right?

LAUREN   Not personally, but yes. Elon Musk.

MARK   What about the CFO?

LAUREN   No.

MARK   How many people sit on their board?

LAUREN   No idea.

MARK   That's okay. Most people don't. In how many countries
            does Tesla sell cars?

LAUREN   I don't know that either.

MARK   Where are their products manufactured?

LAUREN   All over.

MARK   Can you name where they have their plants?

LAUREN   Some. But not all of them, no.

MARK   What are their competitors working on to take them out?

LAUREN   That we know. There are a bunch of EV car makers.

MARK   Great! That's right. There are a bunch of EV car makers, but what specific improvements in R&D are they working on?

LAUREN   I have no idea what they're doing behind the scenes.

MARK   Don't worry. You're doing great. How do you feel right now?

LAUREN   Good! I feel like I'm learning something.

MARK   So, we can keep going?

LAUREN   Fire away.

MARK   That's the spirit. So, how many employees do they have?

LAUREN   I have no idea.

MARK   Tesla doesn't have traditional dealerships, but do you know how many locations there are where you can buy their cars?

LAUREN   A lot.

MARK   But do you know specifically how many?

LAUREN   Not exactly.

MARK   How many people bought electric cars last month?

LAUREN   Quite a few.

MARK (laughing)   Yeah, but—

LAUREN (good-heartedly)   No, I don't know how many.

MARK   What was their net income after taxes last year?

LAUREN   Don't know.

MARK   Don't worry. The vast majority of the people who own Tesla don't know either. What were their gross sales last quarter?

LAUREN   Don't know.

MARK   Now, let's really get into the weeds. What's the book value of the company?

LAUREN   Don't know.

MARK   What's the company's price-earnings ratio?

LAUREN   No clue.

MARK   Right. Why would the average investor know that? But what makes this hard for the professionals to know is that even some of this so-called known information is fuzzy. Statements on balance sheets might not be accurately put together. Do you know if Tesla uses LIFO, FIFO, or average cost accounting for their products because the form of accounting they use determines their reported profits.

LAUREN   I do not.

MARK   These facts and statistics determine the company's success, which determines the stock price. There are armies of people at all of the big brokerage firms and mutual fund complexes who study everything there is to know about these companies, and they're getting paid millions of dollars a year to do it. Do you think you know more than they do about the real price of that stock?

LAUREN   Of course not.

MARK   Perfect answer. Nobody does. It would be like you trying to win a swim meet against Michael Phelps. This is not a fair fight. These people spend their entire career learning as much as possible about a single company like Tesla.

LAUREN   I think I'm starting to get it.

MARK   So, who do you think knows more about the company than anyone else in the world?

LAUREN   Elon.

MARK   That's right. So, is it fair to say that you definitely don't know more about the company than him?

LAUREN   That's fair.

MARK   And have you ever seen a CEO try to predict their dividends, just for the next quarter? It's only three months out, and they are constantly wrong. When those predictions don't pan out, they'll go on 24-hour cable news to be grilled by the anchor about why they were so wrong. Do you know what they say?

LAUREN   No, but I have a feeling you will tell me.

MARK   I am. They'll say, "There were some things that happened that we didn't expect." Even the CEO gets this information wrong, but let's say there was some information out there that would help determine the price of this stock. Do you know anything about this company that nobody else knows?

LAUREN   Absolutely not. Only what I've read or seen on TV.

MARK   But by picking this one stock over thousands of other stocks you're making an unspoken theoretical assertion. That's why we call this a thought experiment. It's to test out the assertion that by picking this one stock over thousands of others, you believe that it will do better than the others, or that it might even be the best stock. Or else, why pick it? That's what stock picking is all about. It's the illusion that you know the best stocks in advance to the exclusion of other stocks.

LAUREN   Well, I'm not doing it intentionally.

MARK   I know. Nobody does. That's what makes this such a conundrum. We rarely think through the logical implications of our actions.

LAUREN   At least I'm in good company.

MARK   You are, indeed (laughing). So, you know a few things about this company, but what percentage of all the possible knowable information about Tesla and their competitors do you have?

LAUREN   Very little.

MARK   Perfect answer, but what percent?

LAUREN   Maybe 1%.

MARK   Do you think it would be that high?

LAUREN   How about a ½%?

MARK   It's probably much less than that, but I'll be generous. And that's only the *knowable* information. Now, you want to consider the *unknowable* information. This includes the biases, emotions, instincts, and the plans of all the people who own Tesla stock. For example, let's say someone suddenly sells $3 million worth of Tesla stock so they can buy a house in Bocca. That will increase supply and cause the price to fluctuate. You can't possibly know that about this individual, or the infinite combination of biases, instincts, emotions, and plans of the millions of people whose actions ultimately influence the supply and demand of that stock. Can you see how absurd it is for anyone to predict the behavior of 8.1 billion people?

LAUREN   Yes.

MARK   And what do we call all of that information that's happening 24/7 on TV and the internet?

LAUREN   The news?

MARK   Right! The news! And by definition, *news* is "new" information. It's right there in the word. And that can change the price of the stock. You don't have a crystal ball that will tell you the news that will happen in the future, do you?

LAUREN   Not yet.

MARK   When you find it, be sure to get me one, too (laughing). And you don't have any insider information?

LAUREN   No.

MARK   And if we haven't already driven a truck through the belief that you can pick the best stocks, did you realize that there are 2,239 stocks that did better than Tesla over the past 12 months?

LAUREN   No! I actually thought it was one of the best.

MARK   Do you have any idea why you didn't pick any one of those other stocks?

LAUREN   (joking) They aren't going to Mars.

MARK   Right. The short answer is that you didn't know what the best stocks were going to be. And you also had a story that involved what Elon Musk was doing with SpaceX, which actually has nothing to do with Tesla and Tesla stock. They are two completely different companies.

LAUREN   That's true.

MARK   If you or anybody else could truly pick stocks, you'd pick the stock that performed the best every year, but nobody can do that. In reality, we have so little information about the companies we invest in that we can't possibly make a decision based on science. It's all based on bias and a story we tell ourselves. Given what you know now, can you see that it's absurd to stock pick, market time, and track record invest with Tesla, or any other stock?

LAUREN   I do.

MARK   It's inevitable because you know a small fraction of what's knowable about a stock, and you know *absolutely nothing* that is unknowable, such as bias and the plans of 8.1 billion people. And it's that unknowable information that will determine supply and demand and the future price. On top of all that, the knowable information and the unknowable information will all change based on new news. So, we are agreed there is no way you can possibly predict the price of Tesla stock based on information you have, and there is no possible way to predict the future?

LAUREN   Yeah, and it makes me feel like I should sell it right away.

MARK   Obviously, now you can see you have been speculating and gambling with your money?

LAUREN   Yes.

MARK   Can I ask what is your true purpose for money?

LAUREN   To provide for my family and leave a lasting legacy we can all be proud of.

MARK   Good. How does speculating and gambling align with your true purpose for money?

LAUREN   It doesn't. If anything, it goes against it.

MARK   Do you see how totally absurd that is?

LAUREN   Absolutely.

MARK   Everyone, please give Lauren a round of applause and acknowledge her hard work up here. She gave all of you a great gift by going through that exercise. How many of you experienced a similar breakthrough?

My trainers and I have conducted this thought experiment over 100 times, and everyone eventually has the same epiphany: *It is absurd to gamble with my money—the money I need for my purpose in life and American Dream.*

My epiphany on this subject occurred back in 1991 when I heard the debate between Rex Sinquefield and David Yacktman. It was the first time I learned that all known information about a given stock is already built into the price. Therefore, only new and unknown information can affect the price of a stock going forward. I've watched that theory be proven again and again over the past 30 years.

If there was one unknown that nobody could have predicted in recent history, it was COVID. That turned the world upside down and led to a string of bankruptcies, including Pier One Imports (540 stores gone), Neiman Marcus (34 locations closed), JC-Penny (149 stores closed), Hertz (16,000 employees gone), Cirque du Soleil (3,500 employees gone), and Pizza Hut (300 locations closed). However, you don't need a massive, global-scale event such as COVID to affect the price of a stock. There are world events and changes within an industry that affect supply, demand, and price.

This should be a moment of catharsis, and if that thought experiment didn't completely destroy the belief that you have the knowledge required to pick the best stocks, then nothing else will. Everything from this point forward will depend on you making this final discovery, because if you get this one down, you automatically get the others. Write it down and repeat it out loud: *"It is abjectly absurd to speculate and gamble with my money."*

****

The self-realization that comes from these five discoveries is the first step to investing peace of mind. But as I said from the beginning, information alone won't help you. Even though I'm asking you to write down these discoveries and repeat the words out loud, it doesn't matter what you write or say. It matters what you truly believe. Your behavior and decisions will be based on your beliefs, and you need to be 100% certain that you find stock picking, track record investing, and market timing completely absurd so you can truly eliminate it forever. This isn't something you can "sort of" be on board with. You must believe it to your core. If you don't, there is a very good chance you will repeat those mistakes in the future, or you will slip back into familiar and addictive behaviors when times get tough, or worse, you experience an amygdala hijacking. Times will get tough, and you will be tempted. You won't be able to stop it from occurring, but when you can recognize it, you can interrupt the pattern.

You will know for certain the moment you make these discoveries, because when you discover something, it will knock you back. It will stun you. It's like opening up your refrigerator and seeing the Grand Canyon inside. I experienced this first when I went from a world of no God to a world with God, and I experienced it again when I was first introduced to the efficient market hypothesis during that debate in 1991. If what you've learned doesn't blow you back, then you didn't discover it. But once you do make this discovery, nobody will ever be able to take that knowledge away from you.

This all sounds great on paper, but the truth is that it is extremely difficult for any individual to do on their own. To be successful and live an extraordinary life, you must have a process and follow the strategies. It begins by asking questions and embracing the screen of the American Dream. You then want to improve your relationship with money, which involves eliminating the no-talk rule and doing battle with your money demons, so you can establish your true purpose for money. It continues by believing in the five discoveries, so you understand how the investing industry and your own cognitive bias stand in the way of you and your purpose. Only then, can you apply the empirically tested, Nobel Prize–winning academic methodology.

My objective is to ensure that your actions as an investor align with the academic research, and I want you to believe in that research like you believe in gravity. When used properly, this information will fuel your American Dream and lay the groundwork to live an extraordinary life filled with freedom, fulfilment, and love.

If you can understand the five discoveries, you have a good chance of accepting the academic-based methodologies in future chapters. If you fail to fully embrace these discoveries it will be all too easy for the Wall Street bullies to keep you under their spell.

Stock picking, market timing and track record investing are seductive "mojo."

So, ask yourself this question: What are the root causes of my investing mistakes?

Notes

1. Bob Pisani, "Why Market Timing Doesn't Work: S&P500 Is Up 14% This Year, but Just 8 Days Explain the Gains," CNBC, November 8, 2023. https://www.cnbc.com/2023/11/08/market-timing-doesnt-work-sp-500-14percent-rally-explained-by-just-8-days.html.
2. Ibid.
3. CNBC, October 4, 2010.
4. "The Today Show," NBC, October 6, 2009.
5. "Closing Bell," CNBC, 2008–2009.
6. Carla Fried, "Jack Bogle: I Wouldn't Risk Investing Outside the U.S.," Bloomberg, December 15, 2014. https://www.biznews.com/briefs/2014/12/15/jack-bogle-wouldnt-risk-investing-outside-u-s.

PART

Making Your American Dream a Reality

"People are usually forced to change. We don't want to change, and then something absolutely forces us to realize that what we are doing isn't working, or that our picture of the world is wrong. We fail. So, we change."

—*Ira Glass*

The reason I was so captivated by Rex Sinquefield's argument during his 1991 debate was that there was a robust body of theoretical and empirical research supporting his ideas, specifically, the efficient market hypothesis. This is a concept developed by Dr. Eugene Fama that states the current price of a stock reflects all the known information, so any price change can only be attributed to information that is currently unknown. Simply put, only unknown and unpredictable information can change market prices going forward.

Dr. Fama was an economist at the University of Chicago. He was also on the board of a company called Dimensional Fund Advisors (DFA), founded by David Booth and Rex Sinquefield in 1981. Dan Wheeler worked for DFA as a consultant to other advisors like me.

Dimensional used Dr. Fama's research (which would eventually result in his receipt of the Nobel Prize) when working for a select few Fortune 500 pension plans and largely institutional investors. The efficient market hypothesis wasn't widely used by advisors on main street because broker-dealers hated the idea and still do. But Dan believed that everyday investors could benefit from this approach, so DFA allowed him to work with advisors to create a brand-new market among individual investors. It was a radical new business model, and few people believed it would work. When I learned about this, I saw the opportunity to overturn the existing investing paradigm. The idea that advisors could comprehend and teach advanced portfolio theory and economics to investors seemed ludicrous. Selling products that had a hot two- to five-year record to investors seemed like a much easier path. To this day that is mostly true.

Dan Wheeler and DFA had the answers I sought. After watching the debate, I found Dan at a booth in a crowded conference hall, but he wouldn't talk to me. "I don't talk to anyone under 40," he said. "You haven't experienced enough pain and suffering to know that the traditional way of investing doesn't work."

"I don't need to spend another 20 years doing it wrong. I know what I'm doing is garbage. I need you to help me fix it," I told him.

I don't remember exactly what I said to convince him, but he agreed to come to Cincinnati to meet with me and my father. That was the beginning of a lifelong friendship and strategic relationship that would last decades.

I returned home from that conference in San Francisco feeling like Jack in *Jack and the Beanstalk* after he just traded the cow for magic beans. When I told my dad that everything he had been doing for the past 30 years was wrong and doomed to fail, I expected resistance, a lot of resistance. In fact, I thought I might not be able to convince him to give up what we had been doing for a new way of investing. That's not what happened. He saw the same thing that I did. I had already eliminated commissions, and now we both understood that the next step was to eliminate active management. We had to completely revolutionize how we invested our clients' money. That required a tremendous leap of faith.

Although everything made perfect sense academically, and the direction we needed to take our new business was clear, there wasn't any existing business model we could follow when starting our new enterprise. It was exhilarating and frightening simultaneously because there was no proof that what we were attempting could work for a business that catered to individual investors, small business owners, and other entrepreneurs. We would have to create our own software and new marketing. We had to innovate a new world of investing. My initial marketing materials consisted of an overhead projector and a yellow pad. Oh, how things have changed with modern technology. My most powerful marketing tool was my own burning desire to help investors stop gambling with their hard-earned money—money they needed for their dreams and retirement. From my point of view, we had no choice.

CHAPTER

Establish Your Future View

"Change is the law of life. And those who look only to the past or the present are certain to miss the future."

—*John F. Kennedy*

"What new future will you envision for yourself about the American Dream and your true purpose for money?"

When my father moved our family to Cincinnati when I was five, we suddenly became Bengals fans. This was back in the 1970s when the Bengals were usually terrible, but it was still fun to go to the games. We had season tickets, so my father and I would go together. He'd start drinking beer. Me a Diet Coke. And like clockwork, right around the start of the second half, he would get a little fired up. "Ooh, man! I could've hit that guy harder than that." Keep in mind that my father is 5' 6" and never played football in his life, but get a few drinks in him, and he's suddenly a gridiron expert, yelling from the stands at NFL players three times his size.

My father wasn't alone. He had plenty of company, but that's what happens in the stands. The crowd is full of people who get overexcited and yell at the players on the field. That's easy to do because it's safe up there. There is no risk and nothing on the line

See Appendix II for additional information regarding data in this chapter.

for the spectators. They can say whatever they want, and it doesn't affect the game. That's because all of the action happens on the field. That's where people take risks. That's where the real work is being done. Those people on the field are prepared and are out there killing themselves. It's easy to sit back from the comfort of the stands and judge, condemn, and criticize. It's much harder to create and do. Living your purpose and investing work the same way.

The breakthroughs and the progress you hope to experience when investing can only happen when you step out on the field. If you don't take that action, you can never expect to experience your American Dream. This is not a passive endeavor. It's time to put all the information you've learned to use. However, you want to ensure that you're taking the right action in the right direction, which requires breaking free from the past and telling yourself a new story about your American Dream and purpose for money. It requires creating a new screen of investing called academic investing principles. To do that, you must first dismantle the old existing screen of investor prediction syndrome.

Investor Prediction Syndrome

To live an extraordinary life, you must see the world through an empowering screen that enables you to create, change, grow, and seize the opportunities at your feet. The world is a beautiful place full of mystery and wonder. The problem is that too many people fail to even recognize those possibilities because of their screens. They accept or assume the current way they see the world is the "truth," or just the way the world is, and there is no changing it.

The world is the way it is until it's not. In the 1900s, prominent physicist Lord Kelvin said, "Heavier-than-air flying machines are impossible." That's how he saw the world, which made perfect sense at the time. And because he saw the world that way, he would never take the actions required to create the airplane. Why try something you think is impossible? Those actions wouldn't be congruent with his screen. Lord Kelvin was brilliant. However, if I asked, Who would be more likely to invent the airplane: a prominent physicist or two brothers from Dayton, Ohio, who owned a bicycle shop? You would probably say Lord Kelvin. But the screen through which he saw the world made that impossible.

There are no airplanes until there are. There is no internet until there is. There is no artificial chip that helps you communicate with your computer until someone creates one. If your screen doesn't enable you to dream big and strive to create new possibilities, you're doomed to accept the world as it is and has always been. This is true on a large scale, but also on a much smaller personal scale. How we see the world and our place in it is determined by our screens. That will dictate our thoughts, actions, and, ultimately, our results.

When my family moved from West Virginia to Ohio I was supposed to enter the first grade, but instead, I was told I had to repeat kindergarten. So, at five years old, I developed a screen that led me to believe I was stupid. That's how I lived my life. That screen further evolved when I started to get bullied. Then, not only was I stupid but I was also a coward.

My dad was my hero. I knew him to be strong and courageous, so I didn't think he would be proud of a stupid coward. That was traumatic for me as a kid, and because I believed that about myself, that's how I lived my life. And my screens manifested my reality. When I was in third grade, I still struggled to read. This was long before the days of expensive tutors. Instead, my dad locked me in my room every day and told me I couldn't come out until I read for an hour. I resisted it at first, and that went on for the entire summer. By the start of the school year, I loved reading, and was placed in the highest reading group.

Experiencing that transformation was a breakthrough that helped me re-evaluate the stories I told myself. That's when I came up with a strategy to offset those screens: *win at all costs.* I still didn't believe I was brave and macho, but I falsely reasoned that if I could win, people would respect and love me. Everyone loves a winner, so that's how I tried to prove my value. I set out to win at sports, school, life—everything. On the surface, it appeared to be effective. I went on to get As, I stood up to the bullies, I became a good athlete in high school, and even starred in the school musical.

However, I still didn't feel fulfilled or empowered. I certainly didn't have real confidence or peace of mind. I didn't fit in with the rest of my classmates and felt like an outsider. I was alone and disconnected from my peers—always trying to prove myself but coming up short far too often.

The reason this strategy failed was because I was overcompensating for the story that I told myself without changing my screen. My entire strategy derived from a place of failure because I believed I was not enough. This wasn't unique to me. It's something every kid goes through on some level. At some point, we all believe we aren't good enough. We all experience a failure to be. And we all come up with a different strategy to deal with it. Some become the class clown. Others rebel. Some drop out and do drugs. Some get straight As. Despite the short-term benefits, it never works.

None of that surface level success filled the void I felt inside. At my core, I still believed I was stupid and a coward. That was my story, and subconsciously, I was sticking to it. That's the screen through which I saw the world, and I let that screen eat my lunch and rule my life for the next 30 years.

I've come a very long way since I was a kid, but I still struggle with screens. We all do. However, as Viktor Frankl reminds us in *Man's Search for Meaning*, the choice is ours. Frankl tells the story of helping a widower who was overcome with grief after losing his wife, view his suffering through a different screen. Frankl simply asked the man what would have happened if he had died first. Once the man realized that by living, he spared his wife from the suffering he was forced to endure, he was able to change his perspective. No matter how firm your beliefs, you always have the power to look at life from a different perspective and change your screen.

It's important to understand that it's never just one screen that you must contend with. Your worldview is composed of multiple disempowering screens that will occasionally cloud your judgment. However, recognizing those screens can take away their power. You just need to identify the warning signs. For example, whenever you feel anxiety, stress, fear, or experience a breakdown in the relationships you have with others, it's a telltale sign of a dis-empowering screen that is holding you back. This is your opportunity to look for a new screen. Screens stand in the way of growth and healthy relationships. But if you can swap out that disempowering screen for one that opens you up to living in a world filled with freedom, fulfilment, and love, you can experience tremendous rewards. Screens can unlock passion, creativity, and self-expression. More important, those rewards will come naturally because your thoughts and actions will be in alignment with how you see the world. Then, you will be able to create results and a life you might not have previously believed possible. This is true of everything, including investing.

The main reason why investors struggle to accept and apply empirically tested academic investing principles is that those principles don't line up with the way they see the world. More specifically, they conflict with most people's pre-existing screen regarding investing. I call that screen *investor prediction syndrome*, and make no mistake; it is a syndrome. By definition, a syndrome is a group of signs or symptoms that occur together and characterize a particular abnormality or condition. A set of concurrent things, such as emotions or actions, usually form an identifiable pattern.

Investor prediction syndrome is the belief that you need a forecast or a prediction about the future to become a successful investor. That includes looking for the next best thing, chasing performance, or the fear of missing out. This is how most people view investing. It isn't a screen you or anyone else consciously created—it was thrust on you because of the culture in which you were raised. Everyone is looking for a prediction. They are always searching for that one piece of information that might give them that quick fix or help them get slightly ahead.

In my experience there are two types of investors. The first type is looking for an economic or political forecast, so they can predict themselves. The second type is hoping to find the best advisor who can predict for them. It doesn't matter what type you are; if your portfolio depends on a prediction about the future coming true, it's broken.

As with any screen, when you see the world through investor prediction syndrome, you don't realize the screen is even there. The screen leads to and justifies stock picking, market timing, and track record investing. If you doubt how prevalent investor prediction syndrome has become, turn on any financial show on cable news and count how many predictions you hear in a given segment. Look at any article in a financial magazine, read a blog from a guru, or peruse the latest content produced by big financial institutions, and you'll see predictions about the market, inflation, recessions, commodity prices, currency fluctuation, and the growth of GDP, to name only a few.

Investor prediction syndrome isn't reserved exclusively for economic predictions. You'll also find a constant stream of political predictions about will be president, how the Supreme Court will rule on a specific case, who will win a war, or new laws and regulations. These prognosticators are just throwing ideas against the wall to see what sticks.

If you or your advisor knew what would happen with the economy, politics, and the future, you might be able to predict the best stocks going forward. That's what pundits promise on these news programs, which is why they spend so much time predicting what will happen in politics and with the market. Just as a goldfish doesn't know it's wet, we don't realize we have the screen of investor prediction syndrome because it's how we've been programed.

As you've learned, most of these predictions turn out to be wrong, but what's so dangerous about a prediction is that you have no way of knowing it didn't pan out until it's too late. Once your eyes are open and you can identify investor prediction syndrome for what it is, you can't unsee it. Hopefully, the information in Part II has put you on the path to dismantling this screen because that's essential before you can even consider seeing the world through the screen of academic investing principles (see Figure 9.1).

Will you choose science or mythology? The choice is yours, but seeing the distinction and knowing what you know now, hopefully, the spell has been broken, and you are less likely to fall victim to the siren song. Investing doesn't need to be the way you thought it

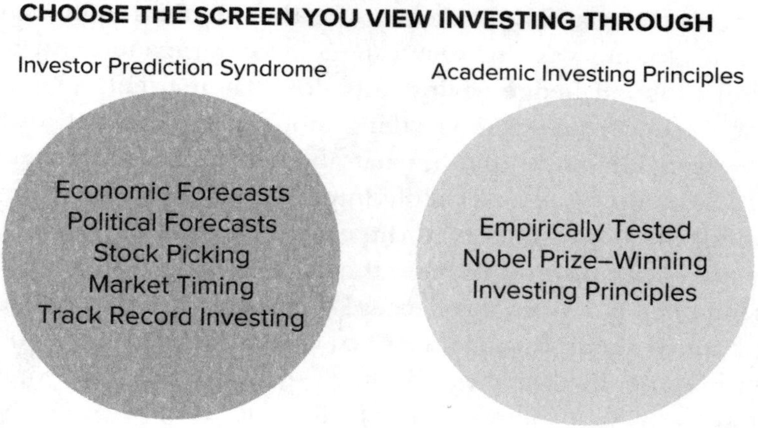

CHOOSE THE SCREEN YOU VIEW INVESTING THROUGH

Investor Prediction Syndrome   Academic Investing Principles

Economic Forecasts
Political Forecasts
Stock Picking
Market Timing
Track Record Investing

Empirically Tested
Nobel Prize–Winning
Investing Principles

Figure 9.1   Two Worlds of Investing

was. When it comes to money, you don't need to live in a world of scarcity and survival. I wrote this book to give you a choice between an ordinary life and an extraordinary life. Once you see the world through the screen of academic investing principles, you will be open to a world of possibilities you never knew existed. It's time to take back control so you can live an extraordinary life of your own creation. That first requires being intentional with your actions.

Intentional Investing

When I go to the grocery store, I aimlessly wander down the aisles, picking things off the shelf that look good until the cart is full. I don't have a list, and if my wife ever sends me to the store to pick up something specific, I have to call her every 10 minutes so she can point me in the right direction. One time, I was in the wrong grocery store. Long story short, my wife doesn't even send me to the store alone anymore.

Many people approach investing the way I do grocery shopping— they aimlessly wander down the aisles of Wall Street without any design or intention. They pick out stocks, ETFs, bonds, Bitcoin, gold, and whatever vehicles they hear are hot at the time. One year, a particular mutual fund looks good, so they get that. Inflation goes up the following year, so they buy gold because some guru told them it's a good hedge against inflation. Their friend keeps talking about Bitcoin, and he's making money, so they want to make money with Bitcoin, too. Oh, what about ESG investing? Let's do that! Maybe they buy a CD while interest rates are high. They just throw all these investments in the cart without much of a plan or even an understanding of how these different investments function together in a portfolio because they haven't thought seriously about any of them. An investor once told me they bought stock in Pfizer because Pfizer makes Viagra. Need I say more? Everyone has their reasons for picking the stocks they do. As you probably discovered in Part II, you didn't pick the stocks in your portfolio; they picked you. Or worse, you paid your advisor to pick them for you. Either way, you weren't investing intentionally. Remember, a good story does not make a good investment.

Unfortunately, this attitude is the natural way of being an investor for most people. And just like the natural way of being a human has us focused on survival, we have the same primary instinct as investors. These investing habits fuel our money demons and give them more power. That leaves us in constant fear as we try to predict the future. It leaves us feeling that money is scarce and that we must always be right so we can win. It leaves us thinking that money is how we measure our self-worth. It's a self-perpetuating cycle of negativity that becomes difficult to break out of because most investors don't recognize that it's even happening. Our returns suffer, and we fail to fulfill our purpose and live our American Dream.

The problem is that your dreams require valid returns, based on a solid foundation of empirically tested academic investing principles. Those principles will help you avoid the paradox of "one single mistake." With the amount of fear, uncertainty, frustration, and confusion surrounding money and investing, these principles are more important than ever. Although I will help you understand how to invest intentionally, I won't force a strategy on you. Instead, I will lay out the academic fundamentals so you can choose your strategy that aligns with your purpose and life.

Before I reveal these principles in Chapter 10, there is one big caveat: you have to use these principles and remain disciplined. That's the tough part, but your dreams are worth fighting for. You can create a new story, and you can be the hero of your own story. No matter what that story is, it begins with how you speak, specifically the language that you use.

Past-Based Versus Future-Based Language

People say that talk is cheap, but talk is the furthest thing from being cheap. When followed up by action, talk is very powerful. Anything that's ever been great or innovative was created first through language before it existed in reality. What you say matters, and you have two options when crafting language: past-based and future-based language. Past-based language describes the current world, or the way it was, while future-based language can create the possibility of a whole new world.

The Wright Brothers were future-based thinkers. They took the necessary actions to create the airplane when brilliant minds such as Lord Kelvin felt what they were attempting was impossible. However, even the Wright Brothers' future-based thinking was limited. Orville Wright said, "No flying machine will ever fly from New York to Paris (because) no known motor can run at the requisite speed for four days without stopping." That's blatant past-based language, which is strange coming from an individual who created a flying machine that nobody thought would ever work. He just couldn't see the next step and got caught in the trap of describing what had been possible up to that time. He's not alone. History is riddled with examples of smart and powerful people held back by their limiting beliefs.

Thomas J. Watson was the chairman and CEO of IBM in 1943 when he said, "I think there is a world market for maybe five computers." Five? I have five computers in my kitchen. My microwave is basically a computer.

Regarding investing, past-based language describes the way things currently are or have always been in the past. And using past-based language is a good sign that you're still seeing the world through the screen of investor prediction syndrome. Here are some common examples of past-based language about investing that I hear all of the time:

- I just need to find the right manager.
- I must learn how to pick the right stocks.
- Buy low and sell high.
- I don't trust anybody.
- Real estate is safer.
- I'll do it myself.
- I already do it this way.
- I succeeded in doing it this way in the past.
- Investing is risky.
- Investing is unpredictable.
- My spouse will take care of it.
- I don't need any help.
- My advisor knows what he's doing.
- Investing is too complicated for me.
- One day, I'll hit it big.
- I have done pretty good.

These are all things you might have told yourself, your spouse, or the advisor you work with. I work with many investors who use this language before even hearing the potential solution. They have no way of knowing if the academic research is legitimate or a better way to do things, and they're already trying to convince themselves that it doesn't work. A quote attributed to Herbert Spencer reads, "There is a principle which is a bar against all information, which is proof against all arguments, and which cannot fail to keep a man in everlasting ignorance—that principle is contempt prior to investigation."[1] In other words, if you haven't seen the research and you're already sure that it's not correct, the screen of investor prediction syndrome continues to plague your judgment.

All past-based language is dangerous, but you must be especially careful if you hear yourself say, "I have always done it this way." That's because "I have always done it this way" is where dreams go to die. Past-based language traps you in the past, and dreams don't have the oxygen to survive in the past. It's like the movie *Ground-hog Day*, when you're doomed to repeat the same actions over and over again. And when it comes to investing, that means continuing to gamble with your money and getting returns that significantly underperform the market.

According to the Dalbar Research[2] referenced in Part II, the average person investing in equities between 1994 and 2023 (30 years) earned an average return of 8.01%. That significantly under-performed the S&P 500 at 10.15% during that same time. Compound that difference over time, and it's a massive loss in wealth, especially when considering inflation.

Investors become so focused on beating the market and assume that they must beat the market to become successful investors, only to come up wildly short. But why try to beat the market when you can own the market? By simply investing in a low-cost S&P 500 index fund on any given year, you historically would have received bet-ter returns than over 80% of actively managed mutual funds.[3] Now consider what those returns look like over 10, 20, and 30 years. Few investors understand that, and in my opinion, that's nothing short of tragic.

Future-based language rewrites the default paradigm originally created by past-based language so you can move beyond investor prediction syndrome and openly accept the screen of academic investing principles. When you change your screen, you will take new actions and get different results. Those results begin with lan-guage, whether spoken, written, or running through your head.

Creating Your Future View

What is the world you want to create—the world where you have adopted the screen of the American Dream, so you can live your purpose? The first step is to identify it. The next step is to become committed to achieving it, but that alone is not enough. You must visualize that future.

Visualization comes easier to some than others, but it's a skill that everyone can develop over time. And it does require work. You want to flesh out that future view with as many vivid details as possible. Those details are important because that's what makes the future world seem real. You want to be able to clearly see this world. When you close your eyes, you want it to appear and play out as if on a movie screen. The best way to begin this process is with a vision statement.

Once I moved beyond my money demon about dating and women wanting me only for my money, I was left with a void. I needed to get clear on what it was I wanted, so I wrote the following statement about what type of person I sought in a partner, and what that relationship would look like:

My wife is a woman with a sense of adventure and fun. She is someone who has a deep passion for life. She knows who she is and possesses a true sense of self-worth. She's secure in her own world and wants a special man to share it with. The woman of my dreams is attractive, energetic, and takes great care of herself and is an athlete. She has awesome respect and honors herself and others. She prefers a man who considers wearing jeans and a T-shirt but can dress up for any five-star event. My wife is not threatened by success and wealth but is also not blinded by it. My perfect mate is intelligent and inquisitive, and she loves to travel with me and has the freedom to take time off and explore. She is dedicated and faithful. My perfect mate wants to be married and is willing to invest the time and energy to build a long-term committed relationship. She enjoys staying at five-star resorts and being pampered but is equally comfortable in a rustic log cabin in a mountain enjoying the outdoors and nature. She loves children and the feeling of belonging to a family. She loves family time together: eating dinner, watching movies, or going on a picnic. She has the power of empathy and understands others' needs as well as her own. She wants a man who is strong and capable of taking care of himself and others. She wants a mature loving partner.

I did this same exact thing in 2014 before building a new company office space in Arizona. It wasn't a simple statement about what

I wanted to accomplish. It was a multiple-page document (that I've included in Appendix I) that detailed every facet of the building, from its look to function.

> The building itself is an experience. There is no front and back-stage—it is all front stage. Think of the space as you would a ride at Disney. It tells a story and the guests become part of that story. . . . The floor changes texture as you move through the space to indicate the next part of the story.

I didn't stop there because a major part of my vision was the impact this building had on visitors, so I made sure to include that in the statement:

> They will feel the power of the community that built this space and understand that they can share in it—they will be part of the community. It will become part of their identity. They will harness the power in their own lives. They will also feel at home and know that this space is created for them. They will feel inspired but also safe and supported. They will feel like they are entering another world. One where the little guy can win—good over evil. They will know that we truly care about them. The space will be fun and convey a sense of adventure.

I worked hard on all of these statements, and fine-turned them until they perfectly encapsulated my vision. I looked at them every day, and it was that repetition that helped my future view become ingrained in my subconscious. It was like I was programming my own brain. This is the thinking behind most advertising campaigns. They show a Lexus driving down Pacific Coast Highway, so you can picture yourself behind the wheel, looking out at that breathtaking view of the ocean. They know that if they can get you to picture that future world, they can get you to act by buying a car. You can use that same tool on yourself with a vision statement that captures your future view of the world.

Despite what it sounds like, this is not magic. It's science. It's how you tap into the brain's reticular activating system to help you see the possibilities and opportunities you didn't see before. It's like if you want to buy a red Corvette, you start seeing red Corvettes everywhere. It's not that they weren't there before. It's that you weren't looking for them, so you didn't see them. You have to know what you're looking for if you ever hope to see it. Opportunities, possibilities, and your dreams work the same exact way.

Now, it's your time to do this, so let's revisit the question from the beginning of the chapter. *What new future will you envision for yourself about the American Dream and your true purpose for money?*

With this statement, you want to take yourself into the future and describe that future world as if it had already happened. Not only create that world in your mind but also reinforce it. Closely examine this new statement to make sure that it's not disempowering or destructive. You want it to be void of all past-based language, so it opens you up to a whole new world of possibilities. You don't want a whiff of anything that could be categorized as investor prediction syndrome in that statement. This new future view should empower you to be creative and take responsibility. It should inspire to you become a leader and a champion of the American Dream who shares it with others. You are the one in control, and you are the hero of your own story.

Notes

1. Alcoholics Anonymous, *Alcoholics Anonymous: The Story of How Many Thousands of Men and Women Have Recovered from Alcoholism*, 4th ed. (Alcoholics Anonymous World Services, 2001), p. 568.
2. "Quantitative Analysis of Investor Behavior," Dalbar, 2023. https://www.dalbar.com/catalog/product/5.
3. Greg Iacurci, "Odds Are, You're Better Off Buying an index Fund. Here's Why," CNBC, March 21, 2022. https://www.cnbc.com/2022/03/21/why-index-funds-are-often-a-better-bet-than-active-funds.html.

CHAPTER

Empirically Tested Academic Investing Principles

"Science is fun. Science is curiosity. We all have natural curiosity. Science is a process of investigating. It's posing questions and coming up with a method. It's delving in."

—*Sally Ride*

"What's holding you back from adopting these empirically tested academic investing principles?"

At this point, you might ask yourself, *If nobody can predict the future, why play the game?* As my dad would say, "Don't throw the baby out with the bath water." I know it's a graphic saying, but it's appropriate here because, contrary to all the problems I've laid out about investing and the industry itself, some good can be extracted if you know how to properly harness the awesome power of free markets.

Equities (also known as stocks) are the greatest wealth creation tool on the planet. For compliance reasons, I have to say that equities are *one* of the greatest tools, but in my opinion, they are *the* greatest wealth creation tool. Stocks are the foundation of capitalism, and your quality of life directly results from capitalism. Think

See Appendix II for additional information regarding data in this chapter.

about it. Your car, home, health care—everything you own—is a function of capitalism. Before a company can create a product, they need to raise *capital*—the root word of capitalism. Those companies need to employ people, so every single job is also a function of capitalism. If there were no capital markets, there would be no jobs. Without a capital market, even your real estate would be valueless. And as long as you don't stock pick, market time, or track record invest, the returns on equities can be phenomenal. Here's an example. Consider these equity asset categories. Here are the annualized returns from 1927 to 2023:

- S&P 500 Index: 10.26%
- US Large Value (Fama/French US Large Value Research Index): 11.93%
- US Micro-Cap (CRSP Deciles 9–10 Index): 11.94%
- US Small Value (Fama/French US Small Value Research Index): 14.30%
- International Large (MSCI EAFE Index (Gross div.)): 9.04% (since 1970)
- International Small (Dimensional International Small Cap Index): 13.47% (since 1970)

These are the market rate returns if you're smart enough to commit to investing long term without speculating and gambling with your money. This is true during the best of times, and it's also true during the worst of times. And there was no worse time in modern history than World War II. It's estimated that more than 80 million people died. Every single major country was negatively affected and experienced loss of life. Europe and Japan were devastated. From 1939 to 1945 was a terrible time for people, but despite all of the death and destruction, the market still prevailed. US large stocks as measured by the S&P 500 earned approximately 12% per year during that terrible time.

Your job as an investor is to achieve market returns without falling victim to the shenanigans of these so-called brilliant people in the industry. The trick is not getting sucked into picking individual stocks or buying into the fantasy that you can beat the market consistently. You have to diversify globally, and you have to be in the market long term. Don't think about the next 20 days—think about the next 20 years. The problem is that most investors are completely in the dark about how markets work.

Understanding the Market

Scroll through Netflix or cable television, and you'll see plenty of shows about space and the cosmos. I love to learn about the universe, so I enjoy those shows. But how many investing shows do you think you'll find that teach the principles covered in this book? You'd be hard-pressed to find a single one. Americans know more about black holes than they do about investing and the market. And I get it. I understand why people love learning about the universe. It's fascinating. However, that knowledge has little impact on your life, your future, your family, and your dreams when compared to what you need to understand to properly invest, yet few people take the time to educate themselves on how markets work.

There are so many misconceptions about money and investing. People tend to oversimplify it and believe that prudent investing is obvious, or at the very most, a limited study of labor, economics, finance, and maybe a little technology thrown in. They check the market on cable news or track their investments online in search of that simple formula, or those hints and tips that will finally provide that one last piece of the puzzle they think is missing before everything clicks into place. In reality, there is no secret. There are no hints and tips because investing is anything but simple. Rigorous science is needed because, in my opinion, investing is the study of everything.

Consider all of the different companies you can invest in. The stock price of those companies is determined by behavior, and behavior is a function of how the human brain works. Human behavior is also determined by the principles of economics, finance, technology, science, philosophy, theology, chemistry, biology, art, law, medicine, design, marketing, music, war psychiatry, religion, geology, anthropology, physics, aerodynamics, history, literature, biology, and artificial intelligence. Human behavior, and therefore finance, is connected to all of those fields, so it is determined, or at least affected, by literally everything. Investing, in a very real sense, is epistemology, or the study of all knowledge. And it's impossible to absorb all of the information required to become an expert in everything. It would take multiple lifetimes to connect all of the pieces. During a time when everyone wants a quick fix, that's not what people want to hear. Remember what happens when you Google the word *investing*.

Early in my career, when I'd tell my clients about academic investing principles, everyone was eager to learn the secret. They wanted to get right to the models, which is probably what you wanted when you first cracked this book. But once I started to explain these theories, and talk about the nature of the market, I'd see their eyes glaze over. To them, I sounded like Charlie Brown's teacher. Honestly, it wasn't their fault. The problem is that markets don't work in ways that are easily understood by human beings.

For many years, I'd look over the research and saw that, despite the short-term volatility, large US stocks had historically averaged about a 10% return per year. That's evident in the yearly returns of the S&P 500. I couldn't help but wonder how the market could be so unpredictable and volatile in the short term, yet so consistent over the long term. I searched, but never found a good answer to that question. After immersing myself for a long time in the subject, after waking up from a nap one afternoon, the explanation came to me out of nowhere: markets are chaotic systems.

Much like the weather, markets are self-regulating. I might know the average temperature in Cincinnati over an entire year is 55 degrees, but who knows what the weather will be like three weeks from today? Nobody has any idea because the weather is a random chaotic system, which makes it unpredictable. They give you a range of outcomes and tell you the probability of it raining because they can't know for sure. Markets work the same way.

Chaos theory postulates that the sum of small effects does not always average themselves out. In chaotic systems small changes can create massive unpredictable outcomes. In this example a butterfly flapping its wings could, in theory, cause a massive typhoon. Much like the butterfly effect, one unrelated and unforeseen event can create a ripple effect that completely changes the price of any given stock at any given time. My youngest sons likened the butterfly effect to a series of dominos. Knock one over, and it can send multiple tendrils of dominos cascading in various directions. One domino can literally topple thousands because of a chain reaction that starts with knocking over a single domino. That's why you can't look at only the financials of a single stock, because it doesn't account for randomness. You can't throw that out and assume the details average themselves out, because they don't. Completely unrelated events can affect the price of a stock leading to random and unpredictable outcomes. Sometimes small inputs can cause the system to go haywire and lead to extreme events. However, in the end, the market will self-regulate. That's how a chaotic system works. It's a far cry from the game or parlor trick that most people assume investing has become. Those who fail to understand this see the world of investing as deterministic (a system with measurable inputs that lead to predictable events) as opposed to being random. Many people struggle with randomness because it means they aren't in control, but quantum physics has gone far in demonstrating the randomness of the universe. In other words, A + B does not always equal C.

Even when you understand the true nature of markets and are open to learning these academic principles, the next hurdle that investors must overcome is complexity. There is no getting around the fact that this material is complicated and challenging. I'd go so far as to say it's almost impossible to understand this unless you have a PhD in statistics. For example, following is the first paragraph from a paper by Dr. Eugene Fama and Kenneth French titled "The Cross Section of Expected Stock Returns":

The asset pricing model of Sharpe (1964), Lintner (1965), and Black (1972) has long shaped the way academics and practitioners think about average returns and risk. The central prediction of the model is that the market portfolio of invested wealth is means-variance efficient in the sense of Markowitz (1959). The efficiency of the market portfolio implies that (a) expected returns on securities are a positive linear function of their market Betas (the slope in the regression of a security's return on the market's returns), and (b) market Betas suffice to describe the cross-section of expected returns.

This is one paragraph. Academic papers on this topic are written this way. How much of it do you understand? Do you know the Sharpe model of 1964, Linter 1965, or Black 1972? Even those who do have a PhD in statistics might be hard-pressed to understand any of this material. Luckily, you don't need to understand all facets of the research to grasp these concepts. Much like you don't need to understand how a plane flies to get on board and arrive safely at your destination.

My objective is to massively simplify the material. So, instead of creating another confusion trap or boring you to absolute death, I want to demystify this material by taking 75 years of academic research and boiling it down to its basic components so anybody can understand it. Most important, I will show you a new way of investing that doesn't rely on speculating and gambling. I will help you build an edifice (the structure or complex system of beliefs) piece by piece using the following three academic principles to help you fulfill your purpose and live an extraordinary life.

1. **Efficient market hypothesis.** Developed by Nobel Laureate Dr. Eugene Fama, this states that current stock prices incorporate all available information and expectations and, therefore, are the best approximation of intrinsic value. Put even simpler: All the knowable and predictable information is already baked into the price of a stock. That means the market is efficient and accurately prices securities. The price is the price. Therefore, only new and unknowable information can change the price moving forward.

2. **Three-factor model.** Created by Dr. Eugene Fama and Kenneth French, this has since become the six-factor model

because the research has expanded. Although all of the risk premiums are critical, I will only go into the three most important ones: market risk, the outperformance of small-cap companies over large-cap companies, and the outperformance of high book-to-market value companies over low book-to-market value companies.

3. **Modern portfolio theory.** Developed by Nobel Laureate Harry Markowitz, this is predicated on the diversification of assets to help you engineer and manage a portfolio that captures those factors from the three-factor model and enables investors to maximize expected returns for the level of risk they're willing to take.

There have been many great economists and Nobel Prize winners who have contributed to the field, but very few of them do research that specifically points to portfolio design and wealth creation. That's what makes the work of Fama, French, and Markowitz so unique and so groundbreaking. Their theories draw on the vast work and body of research of other great economists such as Friedrich Hayek, Milton Friedman, and Thomas Sowell, who helped to demonstrate the efficiencies of free markets, capitalism, and the cost of capital formula.

When you invest based on these three scientific empirically tested academic principles, there is no stock picking, market timing, or track record investing required. This was the foundation of the new approach we took with our clients in 1991. First, we went to every single one of them and said, essentially "Look, we've been doing okay, but the truth is that we haven't been beating the market. And when we did invest your money with managers with great track records, they didn't repeat that performance. We didn't have access to this academic research at that time. Now we do, and we know we can't keep doing what we've been doing."

It's funny that when you admit you've been wrong and show someone a new, improved path forward, they tend to be very understanding. That's why I can't remember losing a single client as we made our transition to the new model.

The foundation of this model is the efficient market hypothesis, so that's where we'll begin.

Efficient Market Hypothesis

There is no greater academic honor than the Nobel Prize, and no other university counts as many Nobel Laureates in the field of economics among its faculty and alumni as the University of Chicago. Among them are giants of finance, such as Friedrich Hayek; champion of free markets, Milton Friedman; and the creator of modern portfolio theory, Harry Markowitz. In 2013, Dr. Eugene Fama joined the ranks when he was awarded the Nobel Prize in Economic Sciences.

Dr. Fama credits his time at the University of Chicago with helping him refine his theories because of the opportunity to workshop those ideas with his esteemed colleagues. While living in Belgium for two years, he wasn't able to receive the same type of feedback, so when he returned to the United States and showed the 10 papers he had written during that period to his colleague, Nobel Prize–winning economist Merton Miller, he discarded close to 70% of what he had worked on because he learned it was junk. That experience taught him that no matter how well regarded he was in the industry, he still wasn't the best judge of his own work. His concepts needed to be reviewed and considered by others. That requires having a thick skin, but it's that scrutiny that helped him produce such groundbreaking work. In 2017, Dr. Fama published a collection of his papers in his massive 814-page tome titled *The Fama Portfolio*, but it was through efficient market hypothesis that I received my introduction to Dr. Fama.

That 1991 debate between Rex Sinquefield and Donald Yacktman was the catalyst moment that would change my perception forever. I had the great privilege of meeting Dr. Eugene Fama in 1992, more than 20 years before he won the Nobel Prize. We've since developed a relationship. He's spoken at my events, and I conducted an extensive interview with him that I still show at all of my workshops. The idea that markets are efficient, and that all knowable information is already priced in, therefore, only unknowable and unpredictable information can change the price became the foundation of my investing philosophy and my business. This is based on the simple premise that free markets work, and the market sets prices better than centralized planning committees. That means the best determinate of pricing is supply and demand. This goes all the way back to what Adam Smith, the father of capitalism, called *the invisible hand*. Because there are trillions of data points that contribute to the price of a stock, no human being, not even the company CEO, knows the true value of that stock better than the free-market pricing system.

Not everyone agrees with the conclusions reached by Fama, Markowitz, and this cohort of economists who believe in the efficiencies of free markets and capitalism. Robert Shiller is another Nobel Prize–winning economist, and a contemporary of Fama, who argues that markets can be overvalued and undervalued. In other words, it's the public's optimism and pessimism that lead to the mispricing of stocks. But after working with investors and studying the market for four decades, I strongly agree with Fama, Markowitz, and their vast body of research that supports the efficient market hypothesis.

Think of it this way: if markets didn't work and stocks were consistently mispriced, smart managers could routinely manipulate the market to make huge returns. Those managers could recognize and sell the overpriced stocks, and clearly see and buy the underpriced stocks. They would be able to repeat that over and over again. Academics have found no evidence of this in the data. And Shiller has offered no practical way to apply his research to effectively extract returns free of risk. Game, set, and match.

Given the large number of money managers, it's expected that some will get lucky and beat the market. That's inevitable. However, as you learned in part two, the managers who beat the market one year, rarely repeat their performance the following year. According to the 2023 SPIVA Scorecard over a 10-year period, 91.4% of active U.S. equity fund managers fail to deliver market returns. If markets were not efficient, you would see the opposite trend.

Even if the ability to beat the market was blind random luck, like flipping a coin, you would expect roughly half the managers to beat the market, but that doesn't happen. Why not? One reason that doesn't occur is because of all the fees and commissions you must pay active managers. When you subtract all of the added costs, those managers will need to overperform the market just to recoup their costs, which can typically be as high as 1% to 2%and in some cases even higher.

So, what about those handful of managers who do outperform the market? It's true, if you could invest your money with them, you would make money. The problem is that you have no idea which managers will fall into that category ahead of time because there is zero correlation between a manager's ability to beat the market in the past and do it again in the future. Nobody knows what those big stocks that make 50% to 100% a year will be. If someone did have that information, they would use it themselves because there is no benefit in telling the masses. But trying to find that manager who will beat the market could cost you in returns, anxiety, stress, and frustration.

All of the evidence supports that only new and unpredictable information and events will change stock prices going forward. Having made the five discoveries in Chapter 8, and truly believing to your core that stock picking, market timing, and track record investing are destructive behaviors, you are already on board with efficient market theory. If the average investor understood this as well, there would be no need for active managers. When I asked Dr. Fama when investors should use active management, his response was, "After they die." Markets are random and unpredictable. Period.

The Three-Factor Model

When analyzing the market, 15 or 20 years is not a large enough sample size. You want a big enough body of data to have statistical relevance. In Fama and French's original study published in 1992, they looked at data from 1963 to 1989 and then looked at data all the way back to 1926 in the US market—some of that data needed to be collected by hand. That's a large sample size, so when you see the same conclusions show up in US markets, international markets, and emerging markets, it further solidifies their findings.

These findings that make up the three-factor model are complicated, so let's begin by explaining a simple economic formula at the root.

$$\text{Cost of Capital} = \text{Return on Capital}$$

Here's what this means: if you own a company and want to raise money to grow your business, you need funding either in the form of stocks or bonds. The riskier the company, the higher the cost of capital and the more you will have to pay your investors.

Here's another way to think about it: if there were two companies—one risky and one safe—and you went to the bank to borrow money to invest in both, which one would have the higher interest rate? Risky companies have a higher cost of capital to raise money, which is why investors have a higher expected return. Stable, strong, and safe companies (like Apple) have a lower cost of capital and a lower expected return. This is counterintuitive for most investors because most of the portfolios I analyze are filled with strong, safe, big company stocks that have a lower expected premium for equities. Years ago, when Apple was a startup, they had a high cost of capital, but they're a growth company now, so they have a lower cost of capital and a lower expected return. Keep this in mind as we dig into each of the three factors because this explains where returns come from if they don't come from brilliant managers picking the best stocks. The returns come from the market mechanism pricing for risk. Don't fight forces, use them.

Factor 1: The Equity Premium

This is the first and easiest to understand of the three premiums, and it's based on the idea that there is a premium for investing in stocks over bonds or fixed income. In other words, stocks are riskier than bonds and, therefore, have a higher expected rate of return, giving investors a premium for owning stocks over bonds.

Between January 1926 and December 2023, the annualized return of the S&P 500 Index was 10.28% (a total return of 1,455,588%), while T-bills were only 3.26% (a total return of 2,214%), putting the equity premium at 7.02% (total return of 1,453,374%).

According to our research, it shows up 71% of the time over one year, and if we go out 20 years, it shows up 100% of the time. If you invested $100,000 in 1926, there was a $1.4 billion premium for owning equities over fixed income. It's hard to picture what that really means, so Figure 10.1 will help put this into perspective.

The graph looks impressive, but when I learned about the equity premium, I still wanted to make sure that none of these findings were purely random, as any good scientist would do. So I asked Dr. Lyman Ott, who has a PhD in advanced statistics and authored the book *An Introduction to Statistical Methods and Data Analytics,* and sits on our academic board. He sent me back a notebook page with statistics scribbled on it. His conclusion: the probability of the 20-year equity premium being random is <<.001. I hadn't seen those marks before (<<), but it turns out they mean "much, much less" than 1 in a 1,000. When I asked, "how much less than 1,000?" I was told that it was somewhere in the ballpark of 1 in 10 million. In other words, it's not random. Lyman Ott went so far as to say, "The equity premium is ridiculously profound."

Equities simply have a higher cost of capital and a higher expected return than fixed income. This doesn't mean that you

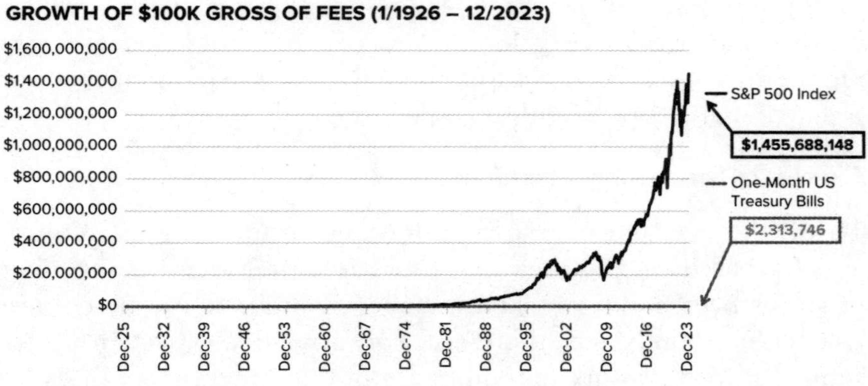

GROWTH OF $100K GROSS OF FEES (1/1926 – 12/2023)

Figure 10.1   S&P 500 Index Versus One-Month T-Bills
Note: See Appendix II for additional information.

want all of your money in equities. It comes down to your risk/reward preference. But even if you are a conservative investor, the equity premium makes a lot of sense for almost everyone.

Factor 2: The Small Premium

This factor examines the often-overlooked benefits of owning small company stocks over large company stocks. Having been in this business for more than four decades and having analyzed over tens of thousands of portfolios from Main Street investors, I know that very few investors have significant exposure to these premiums. I've seen the data, and even those who do have exposure, it's typically a trace amount. Why? Familiarity bias, herding bias, and a number of biases. Whatever the reason, people rarely go out and buy micro-cap stocks. They do what their friends are doing and listen to the gurus on cable news. In Figure 10.2, you can see how the small premium shows up over the long term from 1926 to 2023.

There are periods when the small premium is huge, and there are other times when the trend is reversed, only for it to pop back up again. How will this unfold over the next five years? Nobody knows. That's why you want both small and large stocks. When not investing in both, you're taking on a ton of extra risk without the

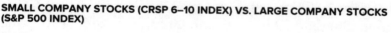

SMALL COMPANY STOCKS (CRSP 6–10 INDEX) VS. LARGE COMPANY STOCKS (S&P 500 INDEX)

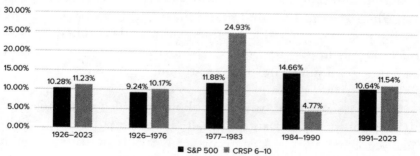

Figure 10.2   Small Company Risk Premium

Note: See Appendix II for additional information.

US SMALL STOCKS (CRSP 6–10 INDEX) VS. US LARGE STOCKS (S&P 500 INDEX)
MONTHLY DATA: JANUARY 1926 – DECEMBER 2023

Rolling Time Periods	1 Year	3 Year	5 Year	10 Year	15 Year	20 Year
Total Periods	1165	1141	1117	1057	997	937
# of Periods Small > Large	596	570	611	662	702	781

Percentage of all rolling time periods in which small outperformed large.

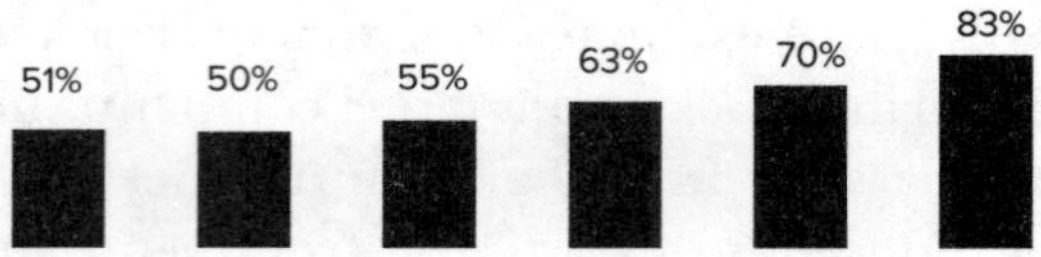

51%   50%   55%   63%   70%   83%

Figure 10.3   Small Risk Premium

Note: See Appendix II for additional information.

same chance for any added return. So, how often does the small risk premium show up over time? See Figure 10.3.

Looking at only one-year periods, US small companies outperform large companies 51% of the time. Over longer periods, the small premium becomes more profound, so when you look at any 20-year period since 1926, small outperforms large 83% of the time. See Figure 10.4 to help put the benefit of the small premium over time in perspective.

If you invested $100,000 in the S&P 500 back in 1926, the overall equity premium you could have earned by the end of 2023 is more than $1.4 billion. If you invest that same amount into only US small companies, your investment grows to over $3.3 billion. That's a massive difference, but is it random? Not according to Lyman Ott. When I asked him that question, he sent me back a similar-looking notebook page filled with equations that reached the same conclusion as he did when I asked him about the randomness of the equity premium. The chances are much, much less than 1 in a 1,000. Again, that's not random.

It makes sense when you think about it. Which companies have a higher cost of capital: large or small? That is another way of asking which companies are riskier to invest in. Small companies must be riskier than larger, stable companies. It's common sense, and if those small company stocks didn't pay a higher return, nobody would ever

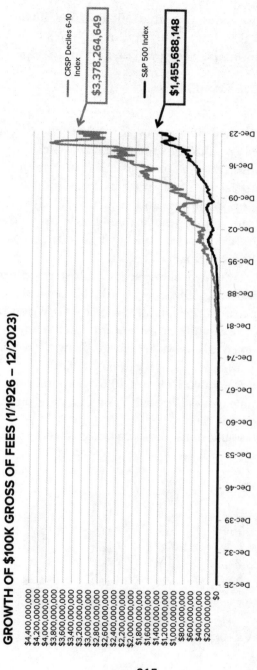

Figure 10.4   US Small Cap (CRSP 6–10 Index) Versus US Large Stocks (S&P 500 Index)

Note: See Appendix II for additional information.

215

invest in them. Just remember that all investing involves risk and unpredictability in the short run. Paradoxically, it's only in the long term that the premiums become evident, and more dependable.

Factor 3: The Value Premium

Let's first define our terms. A value company is a high book-to-market company, which means they are distressed companies compared to the other companies similar in size, assets, and revenue. Distressed companies have a lot of assets (book value) but a very small price in the market. They aren't making a lot of money, they don't have huge opportunities, and you most likely haven't heard of many of these companies. You probably didn't pick them when I asked you to name three stocks, but they have a high cost of capital, therefore, have a higher expected return.

Most people don't own value stocks. Most people have their money in growth stocks, such as Facebook (Meta) and Apple. In my experience, few investors own small value stocks. When people buy small stocks, they are usually growth companies, and small growth companies don't even do as well as large growth companies. The highest expected return is in small value companies.

As you can see in Figures 10.5 and 10.6, the value premium shows up in the United States, internationally, and in emerging markets.

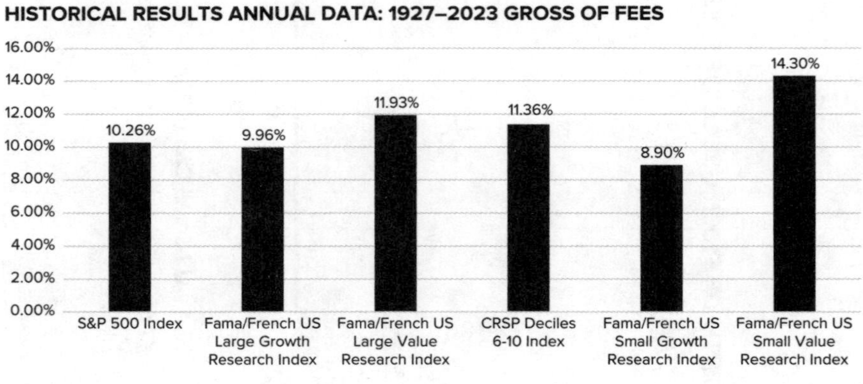

HISTORICAL RESULTS ANNUAL DATA: 1927–2023 GROSS OF FEES

Figure 10.5   Value Versus Growth

Note: See Appendix II for additional information.

As you can see in Figure 10.6, over a 20-year period, the value risk premium shows up 93% of the time. Now compare US large value stocks to the growth of the S&P between 1926 and 2023 in Figure 10.7.

And according to Lyman Ott, the value premium is very dependable over time. There is much, much less than a .001% chance that it's random. Of course, you don't want your portfolio to be made up of all value stocks, but you want some, so you have exposure to this expected risk premium. And you can't just buy into a "value" mutual fund and assume you've got this asset category covered. That's brokerage firm marketing, and it doesn't mean the fund has all value stocks. That is another problem with retail mutual funds. The name of the fund can be very misleading. To understand what is actually in the fund, the underlying holdings must be analyzed and compared to academic asset categories.

****

Put all three factors together and the average annualized return of each from 1927 to 2023 is listed here:

- The equity premium: 8.69%
- The small premium: 2.85%
- The value premium: 4.25%

VALUE STOCKS (FF US LARGE VALUE RESEARCH INDEX) VS. GROWTH STOCKS (S&P 500 INDEX) MONTHLY DATA: JULY 1926 –DECEMBER 2023 GROSS OF FEES

Rolling Time Periods	1 Year	3 Year	5 Year	10 Year	15 Year	20 Year
Total Periods	1159	1135	1111	1051	991	931
# of Periods Value > Growth	641	723	775	811	830	870

Percentage of all rolling time periods in which value outperformed growth.

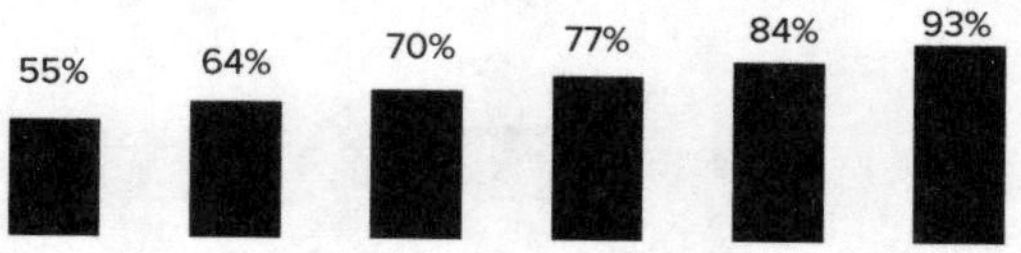

55% 64% 70% 77% 84% 93%

Figure 10.6   Value Risk Premium

Note: See Appendix II for additional information.

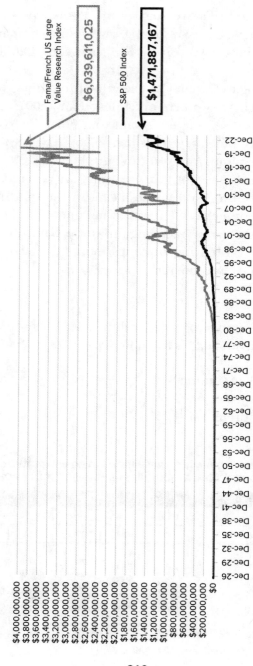

Figure 10.7   Fama/French US Large Value Index Versus US Large Stocks (S&P 500 Index)

Note: See Appendix II for additional information.

To ensure you're capitalizing on these three premiums when investing, you want to consider the three dimensions of stock returns: equity market, company size, and company price. When looking at Figure 10.8, you can see that a portfolio to be in the upper-right corner involves higher risks coupled with higher expected returns. This is exactly what you would expect to find in an efficient market where the cost of capital equals the return on capital. This tells us the type of risk investors are typically rewarded for taking.

You want to be in the upper right quadrant. The problem is that most investors find themselves in the lower left quadrant with large growth stocks where they get the lowest expected return. That's where you end up when you stock pick. However, this research still isn't enough to get some investors on board. That makes this a good time to revisit the question from the beginning of the chapter: *What's holding you back from adopting these proven academic investing principles?*

Yeah, But . . .

I've worked with investors who've looked at this research and believed in academic investing principles, yet still didn't follow through on what they've learned. It's one thing to have the information; it's another to use that information to take action. You

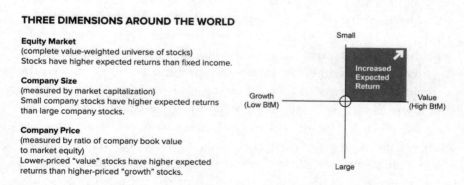

THREE DIMENSIONS AROUND THE WORLD

Equity Market
(complete value-weighted universe of stocks)
Stocks have higher expected returns than fixed income.

Company Size
(measured by market capitalization)
Small company stocks have higher expected returns
than large company stocks.

Company Price
(measured by ratio of company book value
to market equity)
Lower-priced "value" stocks have higher expected
returns than higher-priced "growth" stocks.

Small

Increased
Expected
Return

Growth
(Low BtM)

Value
(High BtM)

Large

Figure 10.8   The Dimensions of Stock Returns

won't get results if you don't get on the field and change behaviors. These investors have come a long way in overcoming their bias and could see the flaws in their old habits but couldn't follow through because of a few lingering screens. I call these the "Yeah, buts," "How about," and "What ifs." In many cases, those who succumb to these biases and screens aren't even aware of them. Here are some common examples, along with an explanation as to why none of these reasons should prevent you from acting.

I want to avoid paying taxes.

I hear this one all the time. "Yeah, but these stocks I own have made gains, so if I sell them now, I will have to pay taxes, and I don't like taxes."

Trust me, nobody hates taxes more than I do. I hate taxes and never want to pay any taxes. However, I'm also aware that there are all kinds of costs for having the wrong investments. Here is a simple example. Let's say you put $500,000 in a stock, and it goes up to $1,000,000. Your gains are $500,000. With long-term capital gains tax about 20% (on the $500,000), that's roughly $100,000 you will have to pay in taxes. However, what happens if the stock goes down? After going up to $1,000,000, that stock could easily drop 40% and erase $400,000 of profits. That's four times worse than the $100,000 you would have had to pay in taxes. Not to mention, you lose all of that money you could have prudently invested just because you didn't want to pay taxes.

My dad used to say, "Use it or lose it." It means that someday when you need that money for your retirement, who's to say that capital gains taxes will remain the same rate as today? To put this in perspective, just remember there were times in recent history when income tax rates were as high as 60% on wealthy individuals. It's easy to forget how much worse it was in the past. Also, there is a very good chance the government could raise capital gains tax or get rid of it altogether. And if you have significant gains in your portfolio, you might be treated as one of those "rich one-percenters," even if you don't consider yourself one. There is also the risk of losing all of the money in the stock. As you've now learned, individual companies do go under, and undiversified investors are often left holding the bag—an empty bag. It's best to lock in that rate now. In other words, take the win.

You must look at tax as being a known cost—20% of the gain. I tell investors to never let taxes be the lone reason why they don't make a prudent investing decision. Taxes are important, but only one consideration when determining a prudent portfolio mix. What's most important is that they find the best possible way to manage their money and then consider taxes. Investors get into all kinds of trouble when they consider taxes first because it can lead to poor decisions.

Cost is important, and I don't want to pay any fees.

There is no free lunch. There is a cost to everything, and always calculate the cost of imprudent investing. If you aren't prudently investing your money according to academic investing principles, your portfolio can take a big hit at any time. In my experience, that is the greatest cost of all.

Fees are important, and must be considered, but they shouldn't discourage you from prudently investing your money. You want to consider how to get the theoretical highest rate of return with the least amount of volatility. You want to engineer a portfolio to maximize the expected return for whatever risk you're willing to take. Not doing that could cost you more than any fees that you might pay to implement an academic approach.

What if I'm already doing pretty well?

Even if you think you're doing "pretty well," you should run a time- and dollar-weighted rate of return report. If you're not comparing your rate of return against a globally diversified, efficient portfolio, then you can't accurately say that you're doing pretty well. And remember that pretty well may not be good enough when your American Dream hangs in the balance. Pretty well is settling and selling yourself short.

I have a cause I want to support with my investments.

Some investors want to invest only in companies that align with their political beliefs and social causes. That might mean not investing in companies that support or fund war, tobacco, or energy. These are all valid views, and it's great that we live in a country where we're free to express our views without fear of persecution. However, when you invest for noneconomic reasons, you frequently and severely limit your returns.

As an investor, your hands are never clean. Here's what I mean by that: let's say that you don't want to invest in a company that supports war. It's easy to avoid purchasing Raytheon stock or that of other defense contractors. But what about companies that produce steel? That steel could be made to build refrigerators or bombers. Companies that produce paper, computers, or electricity might all fund the military in different ways. Think about how difficult this would be to do on a consumer level. If you were strongly anti-tobacco and didn't want to give your business to any company that sold tobacco products, does that mean you couldn't get a prescription filled at CVS or any of the pharmacies that sold tobacco? This approach to investing will have very little impact on these companies and your chosen cause, but it could severely limit your returns and prevent you from properly diversifying. Not to mention that it becomes a nightmare trying to keep track of this.

A much better strategy would involve you embracing free market capitalism and academic investing principles to get the highest expected return for your desired risk level. You can then use that money you earn to directly support the charities and causes you believe in. That will have a significantly larger impact than trying to manipulate your portfolio to fit your social or political agenda.

I want to diversify risk by using multiple advisors.

This is another common one, but using multiple advisors is not how you decrease risk. Given that there are so many con artists, prognosticators, and bullies, the more advisors you take on, the more likely you are to stumble on one of those incompetent or unethical advisors who will take advantage of you. It doesn't reduce the risk. It can actually increase risk and prevent you from rebalancing, so you can never take control of your desired risk/return portfolio design. You'll often have competing investment philosophies with added cost and unintentional overlap because there is a high chance that multiple advisors will be buying and selling the same thing. Instead, find an advisor who meets all of the criteria (meaning they use and only use academic investing principles) and invest with that one person. I'll explain how to do that later in this part. I'll also show you how to investigate the integrity of the advisors you consider trusting with your money.

I'm better off investing in real estate.

I have no problem with real estate, but real estate is not an investment. Real estate is a business. If you start flipping houses or buying apartment complexes to manage them, that's a business, and there is a lot that comes with managing a business. That's not my area of expertise, but if that's what you're interested in, by all means, you should do it. If you're good at it, you will likely have more money to add to your investment portfolio. That makes it even more important to understand these investing principles. I highly recommend having your own business, but it's completely different from investing. It's not an either/or decision. You can, and should, do both at the same time.

I want better protection against inflation.

I get it. Investors are concerned about inflation, especially as I write this today at the start of 2024, but I've seen it get a lot worse. Back in the early- to mid-1970s, I was young and watching my dad do financial planning when interest rates were about 12% and 13%. The volatility of interest rates is just as difficult to predict as equity markets. Inflation fears are used to sell many inflation-hedge investments. Historically, it can definitely get much worse.

People do crazy things to try and hedge against inflation, such as buying gold. As mentioned, you're taking a lot of risk with gold that has absolutely nothing to do with inflation. Commodities, in general, are not a good hedge against inflation. What should you buy instead? When I posed this question to Dr. Fama, he suggested investing in TIPS, or Treasury inflation-protected securities. Yes, there are times when all fixed income is low, but that's why you diversify, and with TIPS, you get payment for the inflation as well as the promised base payment on the securities. Short-term bonds are excellent hedges for inflation, as are five-year government, high-quality corporate-fixed, and global government bonds.

Should I wait to see what the Fed does? Don't they control interest rates and influence the market?

If you look around the world, you'll see that various central banks have done wildly different things since the 2008 recession. In the United States, we've had quantitative easing where the Fed has issued short-term debt and bought long-term debt, which is a neutral activity. However, the European Central Bank didn't do any of that, and neither did Japan. In other words, central banks around the world did wildly different things and ended up with the same term structures. It's completely false to think that the market goes up or down because the Fed manipulates interest rates. Supply and demand control the interest rates, not the Fed. At the most, they have influence at the margins, but all of their activity follows the market. They don't have nearly the power that people attribute to them. According to Nobel Prize winner Eugene Fama, if they have no impact on the short-term borrowing and lending market, they will have no ability to manipulate the stock market.

****

This list can go on and on. These are all just excuses and reasons why you can't be prudent right now. You tell yourself that you might do this *someday*, but you can't do it *right now*. What it really means is that external circumstances, biases, instincts, and emotions are causing you to violate your beliefs. In my experience, all it takes is one "reason," no matter how trivial and unscientific, to throw all of the research and analysis in the trash. I understand. Change is difficult. It can push you into unfamiliar territory and lead to uncomfortable conversations. It's natural to want to go back to doing what you believe to be safe and comfortable, so your mind will come up with all sorts of reasons why now is not the best time to change something in your life. That's your brain trying to protect your ego, but in the long run, all it will do is prevent you from taking action. Remember, the best time to be prudent is right now. Always!

I tell advisors that there are three things that I must see before I even consider changing portfolios. The first is that it must meet the efficient market hypothesis. The second is that it must be grounded in the cost of capital story. And the third is that I heard it from Eugene Fama. Now that you understand markets are efficient, returns come from the cost of capital, and different types of equities have premiums based on risk/return characteristics, how do you apply these principles?

CHAPTER 11

Diversifying with Modern Portfolio Theory

"The chief problem with the individual investor: He or she typically buys when the market is high and thinks it's going to go up, and sells when the market is low, and thinks that it's going to go down."

—*Harry Markowitz*

"What investing behaviors and actions are you committed to stop doing, and what investing behaviors and actions do you declare to start doing in the future?"

There is no such thing as a practice portfolio. You're not preparing for the real show down the road, and you don't get any Mulligans. Your current portfolio is your real portfolio, and every second counts. You want results, not stories about why events didn't work out in your best interest. To truly use all of these principles described in Chapter 10, investors must understand the dimensions of risk and diversification.

Diversification is the investor's best friend. Talk to the average investor, and they will tell you that diversification is good, but the problem is that they don't understand what that means.

See Appendix II for additional information regarding data in this chapter.

227

There are gurus on television saying that you need only five stocks to diversify when it's far more complicated.

Nobody articulates the many complex aspects of diversification as clearly and thoroughly as Harry Markowitz, the creator of modern portfolio theory. When I took an investing class in college, the book we studied devoted only one chapter to Harry Markowitz, which is a total disservice to the man and his work. Still, when I read that one chapter back in 1986, I believed he was right. What he described made total sense to me, and I wasn't the only one his work affected. Markowitz would win the Nobel Prize in 1990.

Harry Markowitz has been one of my heroes since college. After admiring his work from afar for so long, I finally reached out to him in October 2018 to do an interview, and he graciously accepted. I traveled to his office in San Diego, and we spent the entire day together. We became friends, and I asked him to sit on my Academic Advisory board. He remained a member of the board while continuing to write books and make valuable contributions to the investing industry right up until his death in 2023 at the age of 95.

Understanding Risk and Return

Investors tell me all of the time that they don't like volatility. What they mean is that they don't like downside volatility. Everyone loves upside volatility, but it's important to understand that these are two sides of the same coin, and you can't run away from one. All investing involves risk. However, not all risk is created equal. There is plenty of risk that doesn't come with any reward. If you get caught golfing when a thunderstorm hits, you can take one of your metal clubs and go stand on top of the nearest hill. That's risky, but is there any expected positive return? No! You only get rewarded for certain types of risk. I call these *prudent risks*.

What Markowitz discovered, and what his research later proved, was that with prudent diversification and combining assets with different risk/return characteristics, you could actually increase the expected return of your overall portfolio and reduce volatility. Consider Figure 11.1.

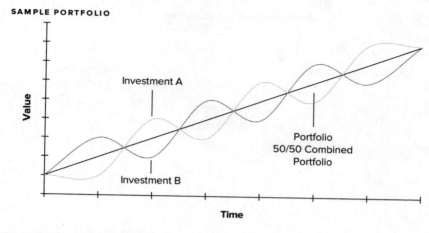

Figure 11.1 Asset Class Correlation

Traditionally, you would think that if you put half your money into security one and half into security two, you would end up in the center line. Markowitz discovered that if you put half in Investment A and half in Investment B, and they're not perfectly correlated (meaning they move in dissimilar ways) then you can actually increase expected the rate of return while reducing risk.

That's the benefit of diversification. Two dissimilar assets, properly allocated, are not as volatile as a single asset. In other words, you can have the same amount of expected return with less risk when you are properly diversified.

If investors could effectively articulate what they truly want (and most can't), the Markowitz efficient frontier pictured in Figure 11.2 would be it. This is how you build efficient portfolios that provide the highest expected rate of return with the lowest amount of volatility. It's what helped Markowitz earn the Nobel Prize in economics. Markowitz called this *mean-variance investing.* Every investor can find a spot on a graph that scientifically shows how much return they can expect for the desired amount of risk they wish to take.

The portfolios with the higher ratio of equities to fixed income (C and D) are more volatile than the portfolios with a lower

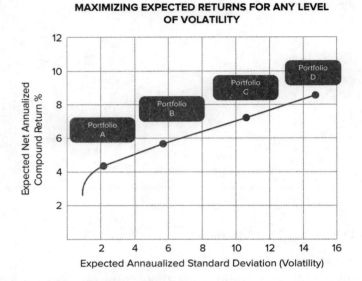

Figure 11.2   Markowitz Efficient Frontier

equity-to-fixed income ratio (A and B). However, it's those more volatile (risky) portfolios that have higher expected returns over an extended period of time.

Volatility is measured by standard deviation, and the higher the standard deviation, the higher the volatility and the higher the expected return for the mix of assets. The idea is to increase the expected return for any given level of risk you're taking. The problem is that most investors have no idea how much risk they're taking on.

You may look at Figure 11.2 and assume you want Portfolio D because it has the highest rate of return over the long term. When the market is performing well, it seems like a no-brainer, but markets change. If you experience a bad year, you might lose for example 40% or more of your money in an aggressive model. The market has already recovered, and over time you are likely to see another increase, but can you live through those rough patches? Can you avoid panic and remain coachable? Can you sleep at night?

There will always be risk in life. If you don't take risks, you're not alive. Risk is part of what makes life vibrant. However, most investors take on more risk than they realize, and their risk tolerance is not nearly as high as they believe it is. It's easy to sit here, look at the facts, calmly consider hypothetical situations, and realize that it's in your best interest to remain on an aggressive path. But when your portfolio goes down, it won't occur in isolation.

Let's imagine you took a very aggressive approach to investing that was heavy on equities because it provided the biggest expected return over time. Things are going great. Then, the market hits a rough patch. That rough patch turns into a rough month. Suddenly, your $1 million-dollar portfolio drops down to $600,000. There's a reason for that. You turn on the news and watch coverage of the latest national disaster. You hear constant reports about inflation and a possible recession. Maybe there's a war. Maybe the housing market collapsed. However, it's not only external circumstances you must deal with. Everyone has personal life circumstances. You might lose your job. A loved one could be diagnosed with cancer. You might have to pay for your child to go to rehab.

During these moments of crisis, rarely will you have to deal with only a single issue. That's a hot state, and there is a very big difference between how you behave in the safety of a hypothetical cold state and how you behave during a very real hot state. In a cold state, you rely on logic, analysis, and planning. You think rationally. But in a hot state, emotions are heightened, causing you to behave impulsively and unpredictably. What makes it so difficult to navigate a hot state is that it can be difficult to recognize when you're in it. You may believe you're behaving rationally when that is far from the case. This is when you're tested, and that's when you will find out who you really are.

It's so important to get out in front of these issues by using your imagination. That involves creating hypothetical situations to help prepare you for what you will do when circumstances go awry. You must put yourself right in the middle of those difficult moments— create both internal and external pressure to see if you can weather the storm. You need to think through these scenarios to prepare. That requires time and effort, but it can make all of the difference when you find yourself in a hot state.

You will be tested, but a great way to help protect yourself from taking unnecessary risk is through diversification. Most investors assume they're diversified but have no way to tell. So, how do you measure if you're diversified or not? Figure 11.3 shows a simple correlation matrix. Yes, it's intimidating at first, but it's something every investor can understand.

One way you don't measure diversification is by the amount of "stuff" you own. You could own 100 different stocks, and multiple mutual funds, but if they're all in the S&P, you have only one asset category. And of the thousands of portfolio's I've analyzed, the majority are weighted heavily toward US large stocks, so there is a good chance that your portfolio is as well.

This graph Figure 11.3 demonstrates which asset categories act in a similar nature. The higher the number the more similar the two assets behave., When you have asset categories with high similarity, it means that if one crashes, the chances are good the other will crash as well. From a diversification perspective, the smaller the correlation, the better diversifier. For example, if you compare US large stocks with international small stocks, you see that those two assets are incredibly good diversifiers. That means you're offsetting the volatility of one asset category with another. However, US large correlated to US large value is .90, which means those asset categories are highly correlated and not as powerful of a diversifier. When U.S. large equities go down, U.S. large value equities will also go down in a similar amount.

You want to diversify among the asset categories that have high expected returns, so they aren't all crashing at the same time. Think of Figure 11.3 as the DNA of diversification. When you examine your portfolio, you want to look at all of the assets you own. Examine the correlations, and then decide what pieces fit together best so you can increase your expected rate of return while reducing your volatility at the same time. It's the closest thing you'll ever get to a free lunch in investing, but, of course, it's not free. It requires study, reading, commitment, discipline, and coaching to maintain your commitment to the process. It won't cost you additional risk.

01/1988 – 12/2023 GROSS OF FEES

	US Large Index	US Large Value	US Small Value	US Micro-Cap	International Large	International Small	International Small Value	International Large Value	Emerging Markets
S&P 500 Index	1.00								
US Large Value	0.90	1.00							
US Small Value	0.78	0.89	1.00						
US Micro-Cap	0.79	0.84	0.96	1.00					
International Large	0.76	0.71	0.64	0.66	1.00				
International Small	0.68	0.68	0.66	0.68	0.92	1.00			
International Small Value	0.65	0.69	0.68	0.66	0.90	0.98	1.00		
International Large Value	0.74	0.75	0.67	0.66	0.98	0.92	0.92	1.00	
Emerging Markets	0.66	0.64	0.61	0.65	0.71	0.70	0.68	0.70	1.00

Color Key

- .8-1.00
- .6-.79
- .4-.60
- .2-.39
- 0-.19
- < 0

Figure 11.3 Correlation Matrix R

Note: See Appendix II for additional information.

It will cost you additional work. In my opinion, that is a price well worth paying.

The Importance of International Diversification

As I mentioned, I believe equities are the greatest wealth-creation tool in the world. But did you realize that nearly 40% of the equity opportunities in the world are outside of the United States? As demonstrated previously, international equities have a dissimilar correlation to US equities, which makes them a good diversifier. International diversification is the key.

There have been periods, like between 1979 and 1989, when international stocks destroyed US stocks. Between 1990 and 2023 that trend reversed and US destroyed international. If you look at 1970 to 2023 as a whole, the S&P came out slightly ahead of international. However, the S&P still wasn't the highest-performing asset category, for it still didn't perform better than international small stocks, and it didn't beat emerging market stocks. And because nobody can tell you what the next 10 years will look like, holding both U.S. and International stocks provides better diversification. Most people don't. Those who do have international stocks typically only have small amounts. And those are mostly large stocks.

There is a world of opportunity in equities. Each country has a different return, and it's impossible to predict how all of these different countries will fare in relation to each other in any given year. For example, in 2004, Austria had the highest return of any developed country, at 71%. Not once has anyone called me up and said, "You know what, I've been analyzing my portfolio, and I think I'm light on Austria." The following year, Canada had the highest return of any country at 28%. In 2013, Finland was at 46%. Since 2002, the United States has had the highest return of all countries only once, and that was in 2014. Every other year, another country posted a higher rate of return, yet whenever I examine portfolios, most are seriously lacking international diversification.

There is no rhyme or reason regarding the performance of international stocks—it's all random and unpredictable, just as it is in the United States. We see the same randomness in emerging market stocks. Emerging markets are countries with less comprehensive trading systems, and the top-performing countries in this asset category can have both huge positive and negative rates of return. Adding emerging markets to your portfolio can increase the expected rate of return, and if you diversify widely, you don't have to take on the full risk of the asset category. By diversifying in many emerging market countries, you can reduce risk. From October 2007 through February 2009, when the US market was tanking (roughly negative 50%), in 2009 Brazil saw a 128% return. Not only do emerging markets have high expected long-term returns but they are also amazing diversifiers. When you compare the correlations from different countries using Figure 11.4, you can see the ones that have the lowest correlation.

Overall, there are over 70 countries with capital markets and more than tens of thousands of unique holdings investors can own. The United States is a great place. You'll find no bigger champion of America than me. Still, I'll also be the first to tell you that you don't want to invest all of your money in just one country. However, it's easy to fall victim to home country bias no matter where you live. Just as most Americans invest in primarily US companies, I've found most Canadians hold primarily Canadian stocks, even though it's a tiny market compared to the US.

Examine the data, and you'll understand why international diversification is crucial. For example, if you correlate Indonesia with Brazil, you're looking at a correlation of .41, which means the countries equity returns are highly uncorrelated. Correlate Malaysia with Columbia, and you have .39, which is even lower. You really want to drive down the correlation because the smaller the number, the more diversified the assets. You do that by adding asset categories with high expected returns and low correlation to the rest of your portfolio.

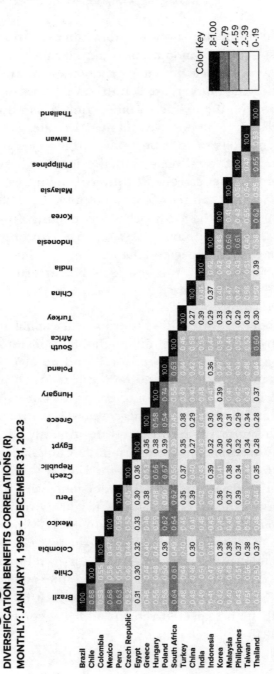

Emerging Market Country Diversification
DIVERSIFICATION BENEFITS CORRELATIONS (R)
MONTHLY: JANUARY 1, 1995 – DECEMBER 31, 2023

Figure 11.4 Emerging Market Country Diversification

Note: See Appendix II for additional information.

The Role of Fixed Income

The high rollers with very aggressive portfolios might have all of their money in equities, but most people don't. They still keep a percentage of their money in fixed income.

Fixed income is the type of investment security that pays interest until its maturity date. At maturity, investors are repaid the principal they had invested plus the interest earned. Government and corporate bonds are the most common types of fixed-income products. Based on modern portfolio theory and research conducted by Eugene Fama, I believe that fixed income's role in a portfolio is to reduce volatility when used in conjunction with equities.

Forty percent of all bonds available on the planet are available in the United States. That's a smaller percentage than equities, and similar to equities, there are fixed-income securities all over the world that provide an incredible opportunity for diversification. And those bonds are made up primarily of government bonds, treasury bonds, and corporate securities.

Make no mistake about it: equities are volatile. And all equity markets are positively correlated, which means when there is a global market crash, all equities are likely to go down to a certain extent. Unfortunately, most people don't realize this until it's too late to offset the correlation. When you look at the S&P 500, you can get some diversification with international large companies, but if you really want to diversify, you want to have short-term, high-quality fixed income because that's how you help protect yourself. That way, when equities crash, your portfolio will likely do better than it would otherwise. With the stability of short term bonds your portfolio may not be as negatively impacted.

Figure 11.5 shows another correlation matrix showing both equities and fixed income; to reduce volatility, it's critical to measure the correlation between the two.

Look at the one-month T-bill correlation to the S&P 500. It's .01That means there is almost zero correlation, which makes it a great diversifier. So, if equities crash, it is unlikely to affect T-bills. You want to have short-term, high-quality fixed income available

Correlation Matrix for Equities and Fixed Income
01/1988–12/2023 GROSS OF FEES

	US Large Index	US Large Value	US Small Value	US Micro Cap	International Large	International Small	International Small Value	International Large Value	Emerging Markets	One Month TBills	Five Year Fixed	1-3 Year Credit	Intermediate Credit
S&P 500 Index	1.00												
US Large Value	0.90	1.00											
US Small Value	0.78	0.89	1.00										
US Micro Cap	0.79	0.84	0.96	1.00									
International Large	0.76	0.71	0.64	0.66	1.00								
International Small	0.68	0.68	0.66	0.68	0.92	1.00							
International Small Value	0.65	0.69	0.68	0.66	0.90	0.98	1.00						
International Large Value	0.74	0.75	0.67	0.66	0.98	0.92	0.92	1.00					
Emerging Markets	0.66	0.64	0.61	0.65	0.71	0.70	0.68	0.70	1.00				
One Month TBills	0.03	0.02	-0.04	-0.04	-0.01	-0.04	-0.03	0.01	0.03	1.00			
Five Year Fixed	-0.02	-0.10	-0.16	-0.17	-0.03	-0.06	-0.08	-0.08	-0.09	0.19	1.00		
1-3 Year Credit	0.28	0.25	0.21	0.20	0.28	0.32	0.31	0.27	0.27	0.34	0.62	1.00	
Intermediate Credit	0.35	0.30	0.24	0.24	0.33	0.35	0.33	0.30	0.29	0.14	0.71	0.92	1.00

Color Key
.8–1.00
.6–.79
.4–.60
.2–.39
0–.19
< 0

Figure 11.5 Correlation Matrix for Equities and Fixed Income

Note: See Appendix II for additional information.

238

to sell and buy more equities when they're down to rebalance the portfolio back to the desired risk/return preference.

Attempting to stock pick is a mistake, but so is bond picking and timing. Many people end up bond picking instead of using modern portfolio theory. Consider that from 1964-2023 one-year U.S. T-notes have a compound annual rate of return of 5.21% and a standard deviation (volatility) of 3.82%. There is some volatility, but it comes with a fairly consistent rate of return. When you go up to five-year U.S. T-notes, you get a higher return at 6.04%, but you double your volatility at 6.57%. But when you go long-term with 20-year US government bonds (which is what most people tend to do in search of higher interest rates), you get a slightly higher return of 6.35%, but the volatility is 12.26%. It's a 1/4% increase in expected return while almost doubling your violability. That's pretty much the same as stocks, except that bonds have a much lower expected return compared to stocks. You might as well just invest in equities, where you can be better rewarded for your risk.

If you're trying to offset the volatility of stocks with fixed income, you don't want to invest in long-term bonds because they are highly volatile and could easily crash at the same time as stocks. Instead, focus on shorter term bonds.

Some people loaded up on bonds prior to 2023 because interest rates were so low. Let's say you owned a $1 million 10-year bond and interest rates start at 2.5% (which is where they were in 2022). If interest rates go up to 5%, then that million-dollar bond is suddenly worth only $785,000. If rates go up to 7% (which is not unheard of), that bond is only worth $650,000. Historically, interest rates were as high as 18%. Let's hope that doesn't happen again, but if it does, you'll get killed if you have long-term bonds. Worse, you'll have much less fixed income to rebalance with.

Interest rates have come down over the last 40 years, but there are still interest rate shocks or sudden rises in interest rates. Every time a shock occurs, long-term government bonds lose significant value. In early 2024, we experienced another rate shock. Long-term government bonds were down 25% at the end of 2023. People who thought they were safe lost a lot of money.

Another potential pitfall when investing in bonds is default risk. Default risk, or the risk that the issuing company or government failing to make its debt or interest payments, is real. They could also default on the full value of the bond. When that happens, it means you don't get your money back. That's why it's crucial to understand the difference between investment-grade and speculative-grade bonds.

Between 1981 and 2023, investment grade or high-quality bonds (AAA-, AA-, or A-rated bonds) had an extremely low default rate of 0.08%. Meanwhile, speculative grade (junk bonds or high-yield bonds that are BB-rated or lower) had a much higher default rate of 3.52%.[1] Brokers love junk bonds. They'll tell you it gives you another 1% or 2% initial rate of return, but what they won't tell you is that you will lose 3% of your money from defaults. And this is the percentage of your portfolio that's supposed to keep it safe. Despite the relatively high default rate, junk bonds are incredibly common.

In 2009, when everyone was fleeing equities for so-called safety, the default rate for junk bonds was 10%. That's the *default* rate. Long-term bonds get hammered during bear markets. If you only look at crashes between 1987 and 2023, the total return of the S&P was −33.85%. If you diversified with high-quality bonds, your fixed income didn't take a big hit because they were up 4.09%. However, not all bonds are created equal—high-yield or junk bonds were *down* −12.15%. That's why owning high-quality short-term bonds helps protect against crashes in the market while making sure you have the liquidity to rebalance during those periods when equities are low.

Rebalancing and Discipline

None of the information in this chapter is worthwhile if you don't have the discipline to rebalance your portfolio. If you don't do that, the makeup of your portfolio will erode over time. It sounds so simple, but few successfully do it on their own. Even the practitioners in the field struggle to rebalance. That's because when you need to do it the most, you want to do it the least.

Here is an easy-to-understand hypothetical. Let's say your port-folio is made up of 50% high-quality fixed income investments and 50% equities. If equities are down and drop below 50%, and the fixed income is up, you sell enough of the fixed income to replen-ish your equities position, at the very time the price of the equity market is down, so you can rebalance your portfolio. When equities are up and over the 50% mark, you sell the excess and move that percentage back to the fixed income. It sounds simple, but when trying to apply this in the real world it is not.

There is a catch. For this to work and to properly rebalance your portfolio, your fixed income must be liquid, discretionary, and available for you to sell without incurring a penalty. People will often say, "My fixed income is this annuity I just bought." That's not liquid—you can't sell it—so you can't rebalance your portfolio. If it's liquid, you can rebalance. It must also be *discretionary*. Because, when the market is dropping, nobody calls their advisors and asks them to sell what's high and buy what's low, at least nobody that I know. Discretionary means that your advisor can balance it for you.

This isn't about trying to predict the future. It's about maintain-ing the same risk/return preference you chose when you created your portfolio. And this is the academic method for doing that. Rebalancing is a systematic method of selling assets as they become relatively high and buying assets when they are relatively low.

The problem is that our human instinct pushes us to engage in the complete opposite behavior. Instincts are very simple. There are two main ones—pain and pleasure. We run away from things that are painful, and toward things that are more pleasurable. Instincts serve a purpose and help us survive. Instincts prevent me from put-ting my hand near a hot stove, so I don't experience the pain. The pursuit of pleasure can motivate me to work hard to provide for my family. However, sometimes the opposite is true. Instincts sometimes drive us toward dysfunctional behavior.

Can you think of something that is painful, but good for you? People will often say exercise. If you're undergoing chemotherapy for cancer or medical treatments that are painful, that might save your life. In those cases, the pain turns out to be good for you. The opposite is true as well. I might find chocolate covered peanuts and other junk food pleasurable to eat, but if listen to my instincts and overindulge, it could lead to health issues. Instincts don't always get it right, and sometimes they get it dead wrong. When it comes to investing, if you listen to your instincts, you can easily destroy a lifetime of wealth creation.

Consider the period between October 2007 and February 2009 (see Figure 11.6). One-year T-notes were up 6.08%, while the S&P 500 was down 50.17%. Yes, over 50% down! If you put $1 million in stocks during that period, your investment was cut in half. If you didn't have the liquid fixed income to rebalance, or worse, you listened to your instincts, you would have sold the stocks when they were low (to avoid the pain of loss) and bought the fixed income when it was up (to run toward pleasure). It might make you feel good at the moment, but you would have taken a huge loss. That's why you want to do the opposite. That's where the opportunity is. Few people can recognize that opportunity, but that is how you rebalance.

The popular strategy of tax loss harvesting completely flies in the face of how to effectively rebalance. Tax loss harvesting is a way people try to lower taxes by selling their investments that

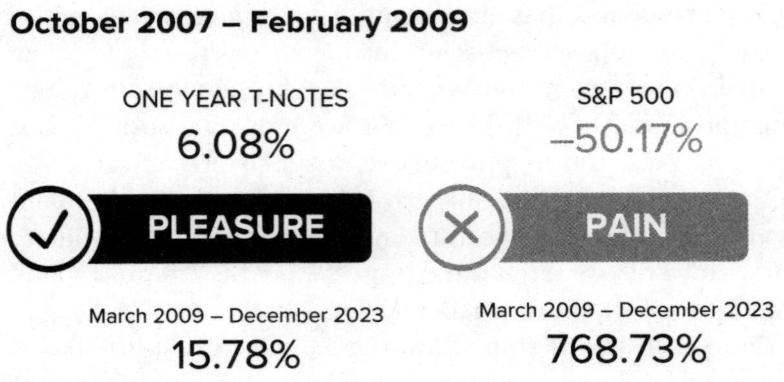

October 2007 – February 2009

ONE YEAR T-NOTES
6.08%

S&P 500
−50.17%

✓ **PLEASURE**   ✗ **PAIN**

March 2009 – December 2023
15.78%

March 2009 – December 2023
768.73%

Figure 11.6 Investing Instinct Diagram
Note: See Appendix II for additional information.

lost money at the end of the year. Yes, it can lower the amount of money you pay in taxes, but it's the complete opposite of an effective long-term rebalancing strategy because you're selling the investment when it's down. You want to sell what's up and buy more of what's down. Also, if you sell an investment when it's down, you have to wait 30 days before reinvesting to avoid paying taxes anyway, so what happens if the market goes up during that period? Now you've lost all of the potential gains because you wanted to save a little on taxes. Tax loss harvesting is the enemy of prudent rebalancing. Keep it simple: sell high and buy low.

What makes successful investing significantly more difficult is that you don't rebalance only once. It's a process that must continuously happen over a lifetime. When people experience a market crash, they often get out of the market and never get back in. That only compounds losses. For example, following when the market was down 50% from October 2007 to February 2009, the market recovered. From March 2009 through December 2021, one-year T-notes were up 11.69%. And that's where you would have put your money if you listened to your instincts and tried to avoid pain and run toward pleasure. However, you would have missed out on the 740.04% gain you could have experienced if you invested in the S&P 500. And this isn't the only time in recent history when this trend occurred.

When the pandemic hit in 2020 and U.S. large stocks were down 30%, did you rejoice and see the opportunity to be had? Did you say, "I just lost 30% of the money I had invested in stocks; it's time to buy more and rebalance!" Don't worry, most people didn't. But that's exactly what I did for clients. It's not a forecast, and it's not a prediction. It's simply rebalancing the portfolio to return to the desired ratio. And guess what happened over the next year? By April of 2021 U.S. Large stocks went up 65%, and U.S. small value stocks, which I've been preaching about, went up 130%. Did I get greedy and advise clients to double down on those small value stocks? No! The portfolio was over its targets. The percentage invested in equities was too high, so I sold the excess equities and put it back into fixed income, so the portfolios stayed balanced.

That's the type of lifelong, iron-clad discipline required to properly rebalance so you can become a successful investor. It may sound obvious and simple now, but this is not easy. I've seen some of the best professionals in the business struggle to remain on the path—it even threatened to strain my personal and professional relationship with my father.

In 1991, when we changed our business model to follow these academic investing principles, everything went very well for the first couple of years. Then, 1995 arrived and over the following five years, tech stocks and US large growth stocks skyrocketed and international stocks, small stocks, and value stocks lagged. These were all trends that seemed to go against our new model, so we had clients calling up and demanding to pull their money if we didn't change our portfolios. Everyone wanted out, including my father. "It's been five years," he told me. "International hasn't performed, small stocks haven't performed, and value stocks haven't performed. I own a big percentage of this company. I want an all-US model, loaded up with large and tech stocks."

"I'm not doing it," I told him. "You can pull your money, but you can't force me to market time. Small, value, and international will have their day. We're going to rebalance and sell US large and restore our allocation back to the original mixes by buying more of the very things you want to sell. I gave my word not to market time and panic, and I intend to keep it."

The argument became heated. That was probably the maddest I've ever seen my father get with me, but he didn't move his clients' money. We both stayed the course during that incredibly long rough patch. Despite the pressure, I never once doubted that I was doing the right thing. I knew that chasing performance was a fool's errand, so when everyone wanted to go all-in on what had been hot for the past five years, I knew it had to be wrong. Buy low and sell high. That's what the research said, and it proved to be correct.

After a five-year stellar US large and tech stock rally, the bottom fell out, and over the next 10 years, we saw a complete reversal. US large stocks and tech stocks had a massive sell off and what resulted is what many people called the *dead decade*, when US large stocks lost nearly 10% over 10 years. Starting in 2000, tech stocks crashed and the NASDQ lost 75%. So much for chasing hot sectors. The system we had in place worked. Everything changed for us, but that wouldn't be the only time I had to do battle with someone I respected.

Dan Wheeler from Dimensional Fund Advisors was my friend, mentor, and one of my heroes. He changed my life when I first approached his booth at the 1991 Schwab event in San Franscico. He was one of the early proponents of this research, and when we first started the company, he went out with me to talk with clients and advisors about this brand-new approach. However, in 1999, he was tempted to create investments that capitalized on recent market movements, specifically to create a large growth index, and even worse, a technology index. I was so flabbergasted that I wrote him a long letter laying out my concerns. Below is an excerpt that encapsulates what I believe the role of an advisor should be. It's something I believed back in 1999, and I believe it more than ever today.

> As always, in the end, you are either part of the solution or part of the problem. Investors and advisors alike chase markets and do not stay disciplined. . . . In my view, promoting—or even allowing—these programs is analogous to a doctor saying, "I don't believe in smoking. It is bad for your health, but my patients like it, and I can profit so I am going to put a cigarette machine in my lobby." It is our job to maintain discipline, not to abandon it for ease or profit. . . . Where there is no discipline there is no true advisor. . . .
>
> First, let's examine the new products introduced at the Advisor Forum, namely the idea of a Growth fund and a Technology fund. In both cases, if recent performance for these asset categories was not so high, I don't think we would even be having this discussion. These possibilities appear to be purely market driven. With regard to the Growth fund, it would be difficult to ignore the massive amount of money sitting in the Vanguard growth fund which many of your advisors possess. Yet, your own research indicates that this asset category has a dependably low expected return and a relatively high standard

deviation. It adds nothing new to the portfolio that is not already represented by the S&P 500. True, there will be years that it will not look like the S&P—so what? This fund only makes sense as a marketing tool. And yet, the possibility of a Technology fund is even worse. Why not a financial fund, or a biotech fund? Why not make a new fund for every asset category that has a good 5-year run?

Shortly after writing this letter, US large companies began a 10-year period when they lost an accumulative 10%. As mentioned, this was called the *dead decade.* Similarly, tech stocks lost over 75% of their value. This is the danger of engineering portfolios based on recent performance and the desires of financial advisors and investors to chase hot markets. For this reason, strict adherence to academic investing principles must be maintained at all times. In the end, Dan's company did not cave into the market pressure to open a technology index. There was no supporting academic reason to do so. In this instance, they held the line, and that was a good thing.

Dan died in 2023 from cancer. He was a great leader with an amazing vision. I owe him a deep debt of gratitude. He was an innovator in the investing field, and he believed in me from the very start. We would not always agree through the years, but we always had a deep respect for each other and a shared commitment to transforming the investing industry. I think about him often, and I still miss him.

There will be plenty of times when you will be tempted to do what you perceive to be safe, convenient, and comfortable. How do you think you will react under those circumstances? It's easy to believe that you'll make the right choice, but the true test comes when times are tough, and conditions are less than ideal. It requires no work at all to adhere to investor prediction syndrome. It's the natural, or default state for investors. It's the water investors swim in. You can predictably stay there your whole life. Getting away from it is where the work will come in. It's a screen that you must fight to replace. That will require you to take a powerful stand and declare your commitments.

The Seven Declarations

When choosing between the screen of academic investing principles and the screen of investor prediction syndrome, the choice is ultimately between a world of confidence and a world of uncertainty—the world of science and data and the world of chaos and confusion. Which world would you rather exist in? Choosing academic investing principles should be a no-brainer. If so, I want you to declare your commitment right now because you get to decide how you invest your money.

A declaration is a formal announcement of the beginning of a state or condition. If something already exists, you don't need to declare anything—you would simply describe the past. And when you describe the past, you can't expect your results to be any different. A declaration uses future-based language to create a world, or some aspect of the world you envision, that does not already exist. There is power in declaration. It's how you speak your purpose into action. Most people believe who they are is a function of their past and view their future as an extension of that past. In the past, you might have done what you were told or what you felt that society determined you should. You only need to declare something if you're trying to change or alter the world and your future. But through declaration, you create the possibility for an entirely new future of your choosing—a future that matters most to you. You are saying that your life is going to be a certain way.

A declaration is an act of responsibility, and once you take responsibility, you have the power to make that future happen. Leave your past in the past, and your future open so you can create from there. That's why the future view is so important. Once you have that future view, you can declare what will happen next. When you make a declaration, you then take actions that align with that declaration. Your future determines who you are in the present. It defines your being.

The world has been changed because of famous declarations. For most of human history, landing on the moon was considered impossible. Before President Kennedy, this was the accepted reality, and that reality was not questioned. Kennedy changed that reality with a declaration:

> *The [United States] should commit itself to achieving the goal, before this decade is out, of landing a man on the Moon and returning him safely to Earth.*

With this declaration, he wasn't describing the past or how things had always been done. He was paving the way for future possibilities that were not likely going to happen otherwise. That declaration set in motion a new way of thinking that led to a different set of actions that eventually changed the world when *Apollo 11* landed on the moon on July 24, 1969.

My favorite and arguably the most famous declaration in American history is the Declaration of Independence. It reads:

> *We hold these truths to be self-evident, that all men are created equal, that they are endowed by their Creator with certain unalienable Rights, that among these are Life, Liberty, and the pursuit of Happiness. . . . And the support of this declaration, with a firm reliance on the protection of Divine Providence, we mutually pledge to each other our lives, our fortunes, and our sacred honor.*

Our forefathers committed their honor, riches, and lives to uphold this declaration. They were willing to die for this declaration, and many of them did. Of the 56 signers, five were captured and tortured as traitors, nine fought and died in the Revolutionary War, two lost their sons in the war, and two had sons who were captured and tortured. Still, they created a possibility for a free new country and way of life through that declaration. They had a vision that was crystal clear in their minds. It made this moment in time possible right now. I wouldn't be where I am today, and I certainly wouldn't be writing this book without that declaration. And you, my friend, wouldn't be reading it, and focused on your purpose and your American Dream. That's the power of declaration.

When I told you that I would help you transform the investing experience to create freedom, fulfillment, and love in your life, that was a declaration. If you are married, you made a declaration that created a new future. When you established your true purpose for money, you made a declaration, and that created the possibility that you could improve life for yourself and your family. Now, it's time to take that one step further.

By making the following seven declarations, you will determine how you will invest your money to fulfill that purpose. Each of these declarations are necessary for you to step into and own a new world of investing. Without these declarations, you have little hope of being able to apply academic investing principles. Think about this as taking a stand or planting a flag. I already gave you the evidence—probably far greater evidence than you had when creating your current portfolio. Now, it's time to declare what you will do to make your future view a reality, just as I did over three decades ago:

- Declaration 1: I declare my commitment to eliminating all speculating and gambling from my investing process.
- Declaration 2: I declare my commitment to investing in the equity premium.
- Declaration 3: I declare my commitment to investing in the small premium.
- Declaration 4: I declare my commitment to investing in the value premium.
- Declaration 5: I declare my commitment to global diversification.
- Declaration 6: I declare my commitment to using high-quality, short-term fixed instruments to help reduce volatility.
- Declaration 7: I declare my commitment to rebalancing and discipline.

Say these out loud. Speak them into reality.

Declarations are not soft, but just because you declare something doesn't mean that it will automatically happen. That's why you can't "sort of" be on board or only agree in theory. By making these declarations, you're stating that you are 100% committed, so go through each one and gut-check yourself. You want to get it flat and down to the bone. You must not proceed without having committed to these declarations. That's how you invest in yourself and bring a new future into existence. It's also how you become 100% responsible for your money and building wealth, so you can live an extraordinary life, free of money demons, self-pity, victim-hood, anger, and regret.

The question you must now ask yourself is if you are currently living up to these declarations. What investing behaviors and actions are you committed to stop doing, and what investing behaviors and actions do you declare to start doing in the future?

It begins with a rigorous portfolio analysis that is designed to identify what's missing from your commitment to the seven declarations. When most investors look at their portfolio under a micro-scope, they discover they did not clearly understand what they owned. So, before we create a road map for you to get where you need to be, let's get flat with where you currently are.

Note

1. "Default, Transition, and Recovery: 2023 Annual Global Corporate Default and Rating Transition Study," S&P Global Ratings, March 28, 2024. https://www.spglobal.com/ratings/en/research/articles/240328-default-transition-and-recovery-2023-annual-global-corporate-default-and-rating-transition-study-13047827.

CHAPTER

12

Where You Are,
Where You Want to Be,
and How to Get There

"Do not conform to the pattern of this world, but be transformed by the renewing of your mind."

—*Romans 12:2*

"What steps will you take to start living an extraordinary life?"

I have loved football ever since I was a kid. Not only did my father take me to see the Cincinnati Bengals but we would also go to all of the local high school games. As early as fifth grade, I knew that I wanted to be captain of the football team. My junior year in high school, I made varsity—I played defensive end and offensive guard. That was one of the best experiences of my life. We finished the season 9–1, and I played alongside a group of dedicated guys. We truly cared about each other. Everyone had each other's back, and the camaraderie we developed when fighting together to achieve a common goal is part of what made us such an elite team.

The following year, I was named captain as a senior, but that experience was much different. The star performers from the previous year had graduated, and the roster was composed mostly of inexperienced underclassman. We went 0–10 and posted the worst record of any Green Hills High School football team ever. I had finally achieved my dream to be the captain of the high school football team, but there was so much anger, resentment, and finger-pointing on that team that there were many moments throughout the year when I wanted to quit. That experience was demoralizing. It was difficult to walk down the hallway at school. And when in high school, everyone tried to find their way and look cool, but I definitely didn't feel cool. I felt like a loser, and at that time, losing was like a slow death. It was the destruction of the life strategy I adopted to comfort myself—to win at all costs. Even worse, if I was a loser, it meant that I would never be a powerful leader. At the time, it felt like my dreams were over.

During our last game of the season, we had the lead going into halftime. It looked like we would finally pull one out. But in the final minutes of the game, the opposing team drove 80 yards down the field to score on the final play and win by one point. When watching that drive, I wanted to throw down my helmet, walk up into the stands, and go home with my parents. I didn't, and I would later learn that decision would make all of the difference in my life.

The season only lasted three months, but at the time it felt like it was never going to end. That was an impactful period for me. I was filled with self-doubt and no amount of positive motivation or attitude could extinguish it. That entire experience made me question my leadership ability, but I never gave up on the team, and I never gave up on being a leader. Years later, I was finally able to look back on that experience and appreciate how having the grit and determination to keep fighting and see the entire season through to the end proved invaluable.

As a business owner and money manager, people constantly question my judgement and my leadership. There are highs and there are lows, but it's during those especially dark times—the dot-com bubble, 9/11, and the real estate crash—when grit and determination are more important than ever. I didn't quit back in high school, and I didn't quit as a professional. To remain on the path, no matter how bleak the situation, is what it takes to be a great investor because you will be tested, and you will be challenged. You will want to quit and do what's easy. How you do one thing in life is how you do everything. When you give in once, that makes it easier to give in again. But staying the course pays dividends of its own. Just remember, you can't live an extraordinary life without taking some hits. You are in this for the long haul, so you must consider the next 10, 20, 30, or 40 years, not only the immediate future. Are you prepared for that future? And when I say the future, I mean all the way to the finish line.

My father used to tell me how he watched so many people work really hard and come so close to success only to let up right before they achieved their dreams, blowing all of those years of hard work, because they didn't finish strong. He called it "nine cents in a phone booth." Granted, this was back when it cost ten cents to make a phone call and we had phone booths, but the concept still applies today. It's about quitting before realizing your vision. It's about fooling yourself into thinking you're already there. To live an extraordinary life, you can't ever let up. You can't stop taking action before you're done. Once you start taking action, make sure you're prepared to run all the way through the finish line. Without that final penny, which might seem trivial, you can't make that phone call.

The 20 Must-Answer Questions

The MRI, or medical resonance imaging, which creates pictures to allow doctors to view the anatomy and physiology of the human body, is a medical miracle. Before the MRI, you often had to resort to exploratory surgery or just live with the pain because X-ray technology was the only way to examine the inner workings of the human body. If you wanted to examine soft tissue, you were out of luck. All of that changed with the MRI. By looking at what's going on inside the body, you can detect problems you wouldn't otherwise notice when looking only at the outside. Your investing portfolio works the same way.

Most investors have what I consider Frankenstein portfolios. These mixes are constructed of disparate parts, thrown together with little true design or scientific method. One year, a certain stock looks good, so it's added. The next year, a mutual fund is hot, so in it goes. Bonds have recently been up, so those are added to the mix. After a few years, investments that looked good in isolation now create a monster resembling the one author Mary Shelley wrote about in her novel, *Frankenstein.* "I have gazed on him while unfinished; he was ugly then, but when those muscles and joints were rendered capable of motion, it became a thing such as even Dante could not have conceived." Like the monster in Shelley's masterpiece, the monster investors create can have disastrous manifestations that were never imagined by their creator. This often happens, even when investors have the best intentions.

When I look beneath the surface to examine portfolios, I find that people have an assortment of everything: stocks, bonds, commodities, crypto, real estate, and a bunch of investments that have fancy and appealing names. Some have maybe 10% in international large stocks, but most investors have no small US, no small value stocks, no emerging markets, and no international small exposure. They are poorly diversified and tend to have a lot of cash because they are afraid of war, recession, and market crashes. Some investors have added index funds and ETFs to their Frankenstein portfolios in the hope that it will be a cure all, but it seldom works that way.

Most investors will go through their entire lives without ever academically examining how they invest their money. They have no clue what's actually in their portfolio or why they own it because there is no intentional design whatsoever. As I said at the very beginning of the book, the quality of your life is directly related to the quality of the questions you ask. And the typical questions and questionnaires that are common throughout the industry are terrible in my opinion. Because the industry doesn't ask the right questions, investors have no idea how much risk they are currently taking on and if they can live with it. And almost nobody understands how their own emotions, biases, and instincts can negatively influence their returns.

As mentioned, peace of mind does not come from portfolio construction. It comes from understanding what you're doing and why at all times. Knowing that, I asked myself: *After working with investors for more than four decades, what are the questions that I ask myself, that make me feel confident that so many other investors never do?* That's how I developed this list of 20 must-answer questions. Go through this list carefully to see which ones you can answer. And for the answer to be a yes, you must be 100% certain, otherwise it's a no.

1. Have you discovered your true purpose for money, that which is more important than money itself?

    If you've done the exercises in Chapter 4, and shared it with someone else, you might already know your true purpose for money. However, you can't just know it, it must be at work in your daily life. If you haven't discovered your purpose yet, that's perfectly fine. For some people, it does take time, but that purpose will be an essential component of your investing philosophy.

2. Are you aware of all the mental biases and blind spots you will likely face as an investor?

    I introduced biases in Chapter 6, and you might have identified a few that apply to you, but there are more than 100, so it would be impressive if you have become familiar with them all. The nature of a blind spot is that, even after you know it's there, you still struggle to recognize yet. And remember, just because you recognize a bias, that doesn't mean you are able to change how it affects you. But awareness is the first step to lasting change.

3. Are you currently invested in the market?

    This is a simple one. Do you own equities or stock-based indexes or funds. Most people can answer yes to this.

4. Do you know three warning signs that you may be speculating with your money instead of prudently investing it?

   If you've read this far into the book, I hope you get this one right. This is a freebie that should be blatantly clear. In case you missed it, the answer is stock picking, market timing, and track record investing.

5. Do you have an academic understanding of how markets work?

   I've provided you with a primer in Part III. How much of it you absorbed, retained, and can apply to your life is a much deeper question, but you should have the basics. Keep in mind that Nobel Prize winner Dr. Eugene Fama's book is more than 800 pages long, so we've only scratched the surface in this book. Some people have spent their entire careers trying to learn how markets work and have made it part of their life mission.

6. Have you defined your investment philosophy?

   There are two basic answers to this question: investor prediction syndrome or academic investing principles. Another way to break this down is to ask if you think markets work or if they fail. If you believe in capitalism and free markets, your philosophy should be academic investing principles.

7. Do you consistently and predictably achieve market returns?

   On the surface, this appears like a simple question, but it's one that many investors improperly calculate because they don't know their true returns. They might think they are getting market returns, or better, but a closer examination of their portfolio reveals that they are falling far short of expectations. Understanding your time-weighted rate of return will go far in helping you identify your actual portfolio performance. Only then can you know the real answer to this question.

8. When it comes to building your investment portfolio, do you know exactly what you are doing and why?

   Even after reading this book, I would highly doubt the answer to this question is a yes. It's just far too complex. However, by getting this far in the book, you are on the right track and starting to ask the right questions—questions that most investors will never ask in their lifetime.

9. Do you have a system to measure portfolio volatility, based on academic research?

   When most people think of volatility, they think in general terms: conservative, moderate, or aggressive. That is not enough. I'm talking about a scientific way to measure the volatility in the portfolio and know what that means. You measure volatility by standard deviation.

   You want to know what your worst possible two-year period could be. Could you lose 10% of your money or 60% of your money? Most people are terrified when I explain to them how much they can lose in a two-year period and had no clue their portfolio was so volatile. So, how do you calculate that?

   Let's say you own one index fund, and, to keep it simple, let's say the standard deviation is 10%, and the average rate of return is also 10%. Roughly 67% of all years will fall between zero (10 minus 10) and 20 (10 plus 10). If you took two standard deviations, it would fall between 30 on the upside and −10 on the downside. And 95% of all the years you can expect to be between those goalposts with only 2.5 years higher and 2.5 years lower. You want to do that for your entire portfolio. It requires understanding basic statistics.

10. Do you have an academic method for measuring your risk tolerance, and do you know what that number is?

   Once you know your current volatility, you then want to know how much risk you're willing to take when constructing your new portfolio so you know how much standard deviation to build into your model. If you have $1 million, can you lose $400,000 and still sleep at night? If not, let's scale it back. Could you lose $200,000 and sleep at night? Time horizon is critical, but the problem is that volatility creates pain and fear that shrinks perceived time horizons, so investors focus more heavily on the short term. That's another bias, because it's one thing to say you can handle a certain level of volatility during a cold state. It's completely different to handle it during a hot state and not panic or become overly optimistic.

11. Have you measured the quality and maturities of the fixed income in your portfolio?

For this, you must analyze the fixed income funds you purchased to better understand the time, volatility, and quality of the bond. Are they corporate bonds or government bonds? If you purchase individual bonds, you want to know the rankings of each and every one. That is a critical step.

12. Do you know where you fall on the Markowitz efficient frontier or how far under it you are?

To determine this, you want to use the correlation matrix to calculate where you are on the efficient frontier. If you're under it, that means you have a mix of assets that aren't getting the expected return for the amount of risk you're taking, or the expected rate of return is too volatile. Rarely, if ever, have I examined a portfolio that is maxed out to the highest expected return for the level of risk they're taking. Most are taking either too much risk or aren't being rewarded for the amount of risk they are taking.

13. Are you working with a financial coach versus a financial planner?

If you're working with someone who educates and trains you to understand these concepts and their applications to your portfolio and your future, that's a coach. If the person you're working with hasn't done that, you're most likely working with a planner. When it comes to investing, a financial planner is more of a black box and an order taker than an educator.

14. Do you have an algorithm and tested process for rebalancing your portfolio, and how often do you apply it?

If you have a way to examine all of the moving pieces in your portfolio, and then use an algorithm or system to apply them on a daily basis, you probably have no reason to read this book because you are eons ahead of the average investor. For most people this is a clear no.

15. Have you measured the total amount of commissions and costs in your portfolio, and their impact on your returns?

     The biggest unknown for many investors is understanding the commissions, spread cost, market impact cost (how much an investor pays over the initial price due to market slippage), and other fees inside the investments they purchase. What's the turnover, or how frequently are the stocks being bought and sold? Is there securities lending going on in a fund, and do the mutual fund companies take revenue from that or give it back? You also have taxes. There are massive implications, and most investors look only at the fee, and not the underlying trading cost that can be very significant inside of a fund. You want to look at everything.

16. Do you fully understand the implications and applications of diversification in your portfolio?

     Diversification is more than just a portfolio filled with a lot of "stuff." Instead, you should consider the correlation matrix from Chapter 11 to put together a portfolio with low correlations in your model, and clearly understand how they all interact with each other.

17. Do you have an investment policy statement?

     This is a written guide that determines how your portfolio will be managed. It incorporates the expected risk, return, and the overall structure of your portfolio. It describes what you plan to do when asset categories exceed their targets. A pension plan at a Fortune 500 company or an endowment plan at a university usually requires an investment policy statement. And having the actual statement isn't good enough; you must understand what it means. The big brokerage firms and your average financial planner don't commonly provide them. I've never encountered a client who had their own investment policy statement when I first met them, but it is critically important.

18. Have you devised a clear-cut method for measuring the success or failure of your portfolio?

Your portfolio should be designed with variance. Hypothetically, if your portfolio were designed with a target rate of 10%, you will not get a consistent 10% return every year. Sometimes it might be 15% and sometimes it might be −5%. You will have variance. The question is how to measure whether your portfolio is inside the designed variance parameters. That variance needs to be understood and articulated in the investment policy statement. You then have to examine the returns each year to ensure they fall within that variance. If you ever fall outside of the variance levels, then you have a problem because your portfolio is not doing what it was designed to do. Someone might see they got a 12% return and think it's good, but it's not if your asset categories saw an 18% return. And just because you lost money, that doesn't mean your portfolio underperformed. If you lost 5%, that's not bad if your maximum downside volatility is 20%. If you don't know how to measure the success or failure of your portfolio, you have no way of knowing if what you're doing is working.

19. Can you identify the cultural messages and personal mindsets about money that destroy your investing peace of mind?

The no-talk rule and money demons are two big hurdles to overcome when regarding your personal mindset. Meanwhile, the cultural messages are like radiation from the sun. They bombard you all day, every day in the form of noise from television, the internet, podcasts, work, friends, and family. You can't escape the beliefs and ideologies that you are exposed to every day. You didn't choose any of them. They were thrust on you and will continue to be thrust on you. Once you are conscious of them, you then have the power to accept or reject them. Awareness is the first step. After that, developing the ability to identify the effect they have on you becomes the objective.

20. Are you ready to shift your personal experience of money and investing from a scarcity mode to an abundance mode—where you can live your life rather than obsess about your assets?

    That last question is the most important of them all because it's a clear signal that you're ready to make a change. This is the only one that you need to answer yes to right now. Most people just want to know what their portfolio mix should be, but as you've learned, it's much more complicated than that. Consider this the first step of a lifelong journey. And the fascinating part of investing is that it will teach you about yourself. You will better learn how your mind works, and what you value.

Don't worry if you aren't able to answer all of these questions. When I do this exercise with advisors for the first time, many of them only get three or four yeses out of 20 questions. And these are professionals who have been financial advisors for decades. But if you've gotten this far in the book, and have been putting in the work, you will be able to answer more of these than most investors. With dedication, and the right coaching, I believe everyone can answer yes to all of these questions. Being able to answer these questions is how you begin to construct a prudent portfolio that can give you the highest expected return for your desired amount of risk. Understanding what you're doing and why at all times is a key determinant to creating lasting peace of mind.

You don't want to invest a nickel of your money until you can answer these questions. That's why I encourage investors to have an independent analysis of their portfolio done, so they can learn these answers. You must arm yourself with knowledge for that is your best weapon when it comes to maximizing your returns so you can live an extraordinary life. Concerning your portfolio creation and management, is there anything more important for your future than maximizing expected returns for the amount of risk you're willing to take?

When it comes to committing to academic investing principles and taking action to align your investing strategy with your purpose, the right time is now. However, this isn't easy. It's complicated, and the stakes are high, so you must go into the investing process with a plan. Boil this down to its bare essentials, and there are three ways to approach investing. The first is to go it alone and do it yourself. The second is to work with an advisor. The third option, which few people know exists, is to work with an advisor who is also a coach. Here is what each plan involves.

Go It Alone and Do It Yourself

Once you've committed to eliminating stock picking, market timing, track record investing, and active management in all its forms, you're already better off than most investors. Theoretically, you could make a portfolio of low-cost index funds and ETFs from big brokerage firms to try and engineer the highest expected rate of return for a given level of risk. There are already many books written about this subject, but it's common to oversimplify this approach because it's not easy to do correctly, and it has its drawbacks. First, let's back up and define some common terms because this is a topic where there is some confusion.

- **Index funds.** This is a type of mutual fund or ETF (exchange traded fund) that seeks to track a market index. This is the performance of a "basket" of securities (like stocks or bonds), which is meant to represent a sector of the stock market, or economy. Its primary objective is to mirror the performance of the index it's tied to. Holdings are changed only at the time of the index's reconstitution, typically on an annual basis. They typically focus on minimizing cost.
- **Structured market funds.** This takes a more dynamic approach and aims to add value beyond the limits of a traditional index. It has a more flexible trading structure and

is focused on capturing market or asset class premiums that deliver value above the index itself because index funds can underperform their actual indexes due to fees, expenses, trading costs, and tracking errors.

Just because you know what a scalpel is and can easily buy one online doesn't make you a brain surgeon. And just because you can easily buy index funds and mutual funds doesn't make you a seasoned practitioner. In the wrong hands, those tools can be ineffective and actually destructive. In addition, index funds can still be used to speculate and gamble. The names of these funds may seem impressive, but most people don't have a clue what's inside them. Investors typically see that a fund has a five-star rating on Morningstar and mistakenly assume that is all the evidence or research they need.

The reality is that you cannot truly determine what's inside a fund going only by its name. The reason is that there are few restrictions on governing mutual funds and what they contain. Even index funds with similar sounding names may contain different underlying assets. That's why you want to examine every single investment in your portfolio, so you can say with absolute certainty what you own. For example, two funds or indexes can have very similar names, such as "small international," but they can each have very different holdings. These brokerage firms don't always define *value* or *small* the same way the academics do, so the returns often suffer as a result.

What also makes the do-it-yourself approach difficult is that many of these crucial asset categories required to diversify your portfolio don't have index funds that you can simply add. This is one area where advisors may have the advantage because they have access to structured market funds and can provide potentially superior returns and discounted fees. Over an extended period, wisely choosing structured funds based on academic research over consumer retail index funds can have a favorable impact on wealth creation.

Figures 12.1 through 12.6 show, respectively, the performance of structured market funds from Dimensional Fund Advisors versus Vanguard index funds for US small cap stocks, US large value stocks, US small value, international large value, international small cap stocks, and emerging markets. In these examples, I have chosen Vanguard index funds because they are some of the most common index funds investors gravitate toward. Compare those returns to the structured market funds offered by Dimensional Fund Advisors, which was cofounded by Rex Sinquefield and where Eugene Fama sits on the board. This firm has embraced the academic investing principles discussed in this book and used them to create their structured funds.

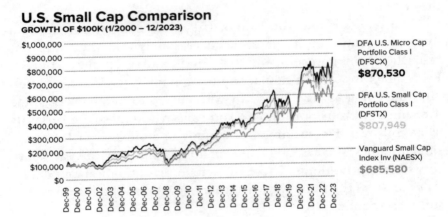

Figure 12.1   US Small Cap Comparison

Note: PAST PERFORMANCE IS NO GUARANTEE OF FUTURE RESULTS. Mutual fund performance information derived from www.Morningstar.com. Fund performance is net of fund expenses. This content is not to be considered investment advice and is not to be relied on as the basis for entering into any transaction or advisory relationship or making any investment decision. Other expenses, such as account-level fees or expenses associated with accessing a fund, if any, are not shown and would lower returns. See Appendix II for additional information.

U.S. Large Value Comparison
GROWTH OF $100K (1/2000 – 12/2023)

Figure 12.2 US Large Value Comparison

Note: PAST PERFORMANCE IS NO GUARANTEE OF FUTURE RESULTS. Mutual fund performance information derived from www.Morningstar.com. Fund performance is net of fund expenses. This content is not to be considered investment advice and is not to be relied on as the basis for entering into any transaction or advisory relationship or making any investment decision. Other expenses, such as account-level fees or expenses associated with accessing a fund, if any, are not shown and would lower returns. See Appendix II for additional information.

U.S. Small Value Comparison
GROWTH OF $100K (1/2000 – 12/2023)

Figure 12.3 US Small Value Comparison

Note: PAST PERFORMANCE IS NO GUARANTEE OF FUTURE RESULTS. Mutual fund performance information derived from www.Morningstar.com. Fund performance is net of fund expenses. This content is not to be considered investment advice and is not to be relied on as the basis for entering into any transaction or advisory relationship or making any investment decision. Other expenses, such as account-level fees or expenses associated with accessing a fund, if any, are not shown and would lower returns. See Appendix II for additional information.

International Large Value Comparison
GROWTH OF $100K (1/200 – 12/2023)

- DFA International Value Portfolio Class I (DFIVX) **$377,833**
- Vanguard International Value Inv (VTRIX) $301,530

Figure 12.4   International Large Value Comparison

Note: PAST PERFORMANCE IS NO GUARANTEE OF FUTURE RESULTS. Mutual fund performance information derived from www.Morningstar.com. Fund performance is net of fund expenses. This content is not to be considered investment advice and is not to be relied on as the basis for entering into any transaction or advisory relationship or making any investment decision. Other expenses, such as account-level fees or expenses associated with accessing a fund, if any, are not shown and would lower returns. See Appendix II for additional information.

International Small Cap Comparison
GROWTH OF $100K (1/2000 – 12/2023)

- DFA International Small Cap Value Portfolio Class I (DISVX) **$720,828**
- DFA International Small Company Portfolio Class I (DFISX) $591,865
- Vanguard International Explorer Inv (VINEX) $374,882

Figure 12.5   International Small Cap Comparison

Note: PAST PERFORMANCE IS NO GUARANTEE OF FUTURE RESULTS. Mutual fund performance information derived from www.Morningstar.com. Fund performance is net of fund expenses. This content is not to be considered investment advice and is not to be relied on as the basis for entering into any transaction or advisory relationship or making any investment decision. Other expenses, such as account-level fees or expenses associated with accessing a fund, if any, are not shown and would lower returns. See Appendix II for additional information.

Figure 12.6 Emerging Markets Comparison

Note: PAST PERFORMANCE IS NO GUARANTEE OF FUTURE RESULTS. Mutual fund performance information derived from www.Morningstar.com. Fund performance is net of fund expenses. This content is not to be considered investment advice and is not to be relied on as the basis for entering into any transaction or advisory relationship or making any investment decision. Other expenses, such as account-level fees or expenses associated with accessing a fund, if any, are not shown and would lower returns. See Appendix II for additional information.

You can see the difference in the returns of the Dimensional funds. They were engineered based on the Nobel Prize–winning research, not the marketing department at big brokerage firms. But if you still believe that you are better off trying to do this on your own, here is another thought experiment called *the problem of infinite portfolios.*

Visualize your potential portfolio as a pie chart. In geometry, a circle has a fixed point in the center, with an infinite number of equidistant points that make up the radius. Mathematically, that means you can have an infinite number of slices. Even if you only have two assets, there is still an infinite number of percentage combinations. In reality, you don't have only two asset options. There are millions of different investment options when building a portfolio, and it's impossible for them all to have the same risk/return relationship. Some will be good, some mediocre, and the vast majority are poor. Very few are superior. Add in the fact that you can trade 24/7, and every second becomes an inflection point when you can change your asset mix. That means the opportunities for errors are also infinite.

Everyone wants to oversimplify investing, but the market is a complex system. And how can you most safely navigate a complex system? You need science to do it effectively. But the science alone won't help you. How do I know that? Because even the people who did the research struggle to remain disciplined.

In his book, *A Random Walk Down Wall Street,* economist Burton Malkiel warned against active management and spoke out against the dangers of trusting the advice of the so-called experts. He believed that because the news was unpredictable, so was the market. His colleague Charles Ellis was another economist who was vocal about the importance of diversification, and he frequently spoke about the dangers of active managers trying to stock pick, market time, and track record invest. He believed the evidence proving it didn't work was overwhelming. An interview with Ellis appeared in one of the early documentaries I created for my workshops.

In 2009, Malkiel and Ellis teamed up to write *The Elements of Investing,* which was intended to be a book in support of their passive do-it-yourself investing strategies that used index funds to remove speculating and gambling from the process. However, in the actual book, they admit that one of them has owned Berkshire Hathaway stock for 35 years, checks the price almost every day, and has no intention to sell, while the other "delights in buying individual stocks and has a significant commitment to China. He enjoys the game of trying to pick winners and believes 'China" is a major story for his grandchildren."[1]

When I first came across that passage, I couldn't believe what I was reading. *Goldilocks and Three Bears* is a story you tell your grandchildren, not gambling in a Communist country that uses slave labor. They add disclaimers to justify their statements by saying their retirement funds are all index funds, and their children only use index funds, but even their attempted explanation is shocking because they also say, "Your odds of success are at least better in the stock market than at the racetrack or gambling casino, and investing in individual stocks can be a lot of fun."[2] In the book, they called this a confession, but a true confession would be admitting how wrong and dangerous this behavior is, and that it was a mistake for them to turn their backs on their own research. Instead, they were gloating about how smart they were.

Apparently, I wasn't the only one shocked and appalled by what I was reading, because in the Foreword of this exact same book, renowned investor, endowment fund manager, philanthropist, and former chief investment officer at Yale University David Swensen wrote, "I cast a skeptical eye on Ellis and Malkiel's explicit endorsement of stock picking in their confessions regarding personal success in security selection. . . . Yes, they can pick stocks—the rest of us cannot."[3] Note Swensen's sarcasm in this direct quote.

It's painful when you watch your heroes go over to the dark side, especially when they were once champions of what you believed. That's why my heart sank when, at the end of my interview with Harry Markowitz, he mentioned his own experience with stock picking. This was the father of modern portfolio theory telling me how he invested in construction materials after learning of recent hurricane destruction in certain parts of the world. I couldn't believe what I was hearing. It was common knowledge that hurricanes had occurred. Based on efficient market hypothesis, that information would have already been included in stock prices. In other words, it was baked into the cake, which makes it abjectly absurd to stock pick. Sometimes, the more you learn and the more successful you are, the stronger the desire to try and game the system becomes, but it's always a trap. If this can happen to the experts who conducted the actual research, what chance does the average investor have of staying the course and remaining disciplined? The problem is that most investors don't understand the difficulty of what they're trying to do.

There are many movies where passengers on a commercial airliner are called on to land a plane when the pilots have been incapacitated. The passengers are almost always successful and make it seem easy in the movies, which is why when men were surveyed and asked if they believed they could successfully land a commercial jet, 50% responded yes.[4] In reality, the number is much smaller. In fact, a passenger on a commercial airline has *never* landed a plane. Furthermore, experienced pilots conclude that this is impossible.

On a lighter note, think of all the people who try out for *American Idol* and are thunderstruck to discover that they cannot sing. Even among those talented people who win these competitions, few become mega superstars. Why? Because it's very hard, and people instinctively underestimate the level of difficulty it requires to be successful at these endeavors. There are just some things you should never try yourself. Never conduct surgery on yourself or a member of your family. I believe investing also fits into this category.

Consider that many economists have dedicated their entire lives to understanding markets and their implications. The Nobel Prize winner Eugene Fama wrote an 800-page book summarizing his greatest work. I've only skimmed the surface of that research in this book. But if you are committed to taking this on yourself, I would start with that 800-page tome. Even if you understand the theories, it is unlikely you could master it yourself. Why? Because this information is available, but it hasn't moved the needle to help the average investor. The portfolios I examine today are no better than the portfolios I examined 10 years ago.

I previously claimed that it was abjectly absurd to speculate and gamble with your money. I also believe that it's abjectly absurd to think you can do any of this on your own. You must admit that when left to your own devices, you are spiritually, emotionally, and intellectually incapable of managing your own money. There is too much noise and too much temptation to successfully manage a prudently diversified portfolio over a long time period. Knowledge does not equal action. That's a mistake I made very early in my career. That's what makes this such a treacherous option, so let's examine option number two.

Work with a Financial Advisor

As an investor, humility is a crucial trait because it prevents you from getting in over your head. Even when you have the best intentions, it's easy to drown in a sea of information. Some people are capable of learning how to rewire their entire house and put the entire electrical system back together. Others wind up burning the entire house down. If you don't want the pressure that comes with investing on your own, the first step is to find a trustworthy and qualified financial advisor. But it can't be just any advisor. You must be absolutely certain that you hire the right person to manage your money, and, as you've learned, not all advisors are created equal. With so many incompetent and unethical advisors, it can be difficult to know where to even begin your search. Whether vetting your current advisor, or searching for a new one, here is a simple four-step checklist to determine if the person you're working with can be a responsible steward of your money.

Work Only with a Fiduciary

This is step number one and can help you avoid many of the problems that run rampant throughout the financial industry. When investing with a fiduciary, you can be confident that your advisor is not trying to put you in models that will allow them to earn a commission whether you make money or not. When you work with a fee-based advisor, you can increase your odds that you both will be sitting on the same side of the table.

Validate the Advisor

First, you want to insist on getting the GIPS (global investment performance standards) of any advisor or institution you plan to work with for the past 5, 10, 20, or 30 years. You want to see what their clients actually earned. Never work with a firm who provides only hypothetical returns or can't provide the returns their clients have made in the past. This way, if you're trying to choose between two advisors, you can compare their track records, and their dispersion, which indicates the variance of returns among all of their clients. However, it's important to note that out of the 32,594 investment advisors that are fiduciaries in the United States, only 3.8% of them claim GIPS compliance.

You also want their income statement, balance sheet, and cash flow statement to be in order and audited by a third party, so you can understand the strength and stability of the firm you may choose to work with. If the advisor can't provide audited financial statements, don't invest with that advisor.

Work with a Third-Party Custodian

You give your advisor the discretion to buy and sell on your behalf, but you want to make sure your advisor doesn't hold the money. That's the job of a third-party custodian. This is how the victims of con men like Bernie Madoff lost all of their money. Even the regulators failed to identify Madoff's fraud. The regulatory bodies didn't verify the assets. Nobody was looking over Madoff's shoulder to make sure he was doing what he said he was doing, so this enabled him to send out fake statements to his clients. When you get your statement from your advisor, it needs to match the statement from the custodian. That will go a long way to help prevent fraud.

Check Violation Tracker

Don't just consider the advisor you work with; consider the morals and ethics of the institution they work for. Violation Tracker is an excellent resource that can help ensure the company you're trusting your money with is stable and has a clean regulatory record. Review the fines and penalties of the companies you work with. This is an excellent way to help protect yourself against fraud and incompetence. Just because you're investing with a big-name brokerage, don't assume they're clean. Look up the violations for any of the big brokerage firms, and you'll find that many have incurred billions of dollars in fines and penalties. Looking at these statistics makes these big institutions look more like organized crime families. Unfortunately, that information is not enough to change the mind of some investors.

An investor once told me, "I don't care about the ethics and morals of the brokerage I invest with, as long as they can make me a lot of money." That is a dangerous way to look at the world.

****

These four steps are crucial, but there remains one big elephant in the room. Investors are human, and so are advisors. The human brain is not hardwired for investing success, and it doesn't matter whether you do this for a living or not.

I have trained thousands of professional advisors in the academic technology of portfolio construction and lifelong ironclad discipline. It saddens me to report that so few of them were able to grasp and implement these strategies in the long term. In fact, most of them caved to their biases, instincts, and emotions, even before their investors felt the same pull. Advisors experience all of the same biases as their clients. The grand renaissance of investing brought on by academic principles never materialized, so many of these advisors are still living in the dark ages. The vast majority of the time, these advisors are stock picking, market timing, and track record investing with their clients' money. In short, they are gambling and thus violating the five discoveries and seven declarations that you so valiantly fought to establish for yourself.

Hiring a financial advisor to manage your money and hope for the best is also an option ripe with danger. The only thing worse than making critical errors with your money is paying someone else to make those mistakes for you. That's why I believe the best possible approach is option number three, and it requires taking this advisor search one crucial step further to make sure the person helping you manage your money can serve another essential function.

Work with an Advisor Who Is Also a Coach

If you look at the data over a long enough time period, you'll see that the market has eventually recovered after a downturn, and the long-term trajectory has always been up. It doesn't matter if it's the

Great Depression or World War II, after every single crash, there is often a fast and furious recovery. This shouldn't be news. Everyone has access to this data—advisors, the media, and investors. I'd go so far as to call it common knowledge. Despite the evidence, whenever the market starts to dip, people will always find a way to panic.

I've thought deeply about this, but what helps me maintain a level head is my belief in the resilience of human beings. We've endured wars, pestilence, floods, earthquakes, and fires, but we always find a way to survive and thrive as a species. We're difficult to kill because it's in our instinct to keep fighting. And just as I believe in the power of the human spirit, I also believe that capitalism, innovation, and creativity are what helps the market bounce back after a crash. Unless we're talking about an extinction-level event where every company across the globe loses all of their value overnight (in which case your portfolio will be the least of your concerns), we know all we have to do is let the market do its job.

If your friend came over to watch the recording of the previous night's football game that you've already seen, you won't be sitting on the edge of your seat, sweating out every play because you already know how this game ends. The same is true about the market, but rarely do we ever think or behave like that. The reason is that, while it's easy to say this now, it's much more difficult to think and act rationally when you start to feel the pressure of everyday life.

In 2021, I was with my family aboard a private plane flying from California to Arizona when one of the engines died. The plane jolted, and the engine grinded to a stop. That's when many of my family members began to panic.

I went up front to speak to the two pilots. They told me that they had to turn off the engine to prevent it from catching fire. We were about 20 minutes away from the closest airport, and that's where we were headed. The pilots appeared calm. They assured me that they practiced these scenarios in a simulator every six months, but as I looked out the window, all I saw were mountains, and

I couldn't help but wonder what would happen if we lost the other engine. I didn't want it to end like this.

Meanwhile, my family was starting to lose it, so I tried to keep them calm. Although I managed to maintain my cool during the incident, I definitely felt a physiological change in my body that I had never experienced before. My heart beat faster, and my mouth dried up. I'd like to think all of my training kicked in and that it didn't show. The only people who weren't panicking were my two young sons, who played video games the entire time. They didn't seem bothered at all. While watching my kids, something hit me. Even though they couldn't fully comprehend the situation, they had the best attitude of anyone on board. Aside from the pilots, there was nothing any of us could do in that situation.

We all held hands and prayed as a family, while the pilots negotiated a landing. The airplane fought through rough turbulence as we closed in on the runway. The airspace was cleared, and the airport closed down for our emergency landing. The wheels touched down and the aircraft bounced multiple times on the tarmac before eventually coming to a full stop. When we exited the plane, we were greeted by a convoy of police cars and firetrucks.

Having worked in the financial industry my entire life, the parallels to investing were obvious. When it comes to the market and your money, it's never the right time to panic, no matter what the circumstances. If anything, panic will only make the situation worse. When you panic, you tend to act irrationally. You make impulsive decisions. You don't think things through. The last thing on your mind is the science of investing. And as you know, one mistake in investing can wipe out a lifetime's accumulation of wealth. Of course, this is easier said than done because panic is a natural biological function. And it's just one of the many reasons that might cause you to make poor decisions.

No matter what you're trying to accomplish, everyone can use a little help. It's a human thing. That support can come in the form of a mentor, a group of like-minded people, or a coach. However, some people resist coaching. It might be their inability to take direction from others, leaving them stuck in a world of "I have always done it like this." Others believe that getting help from a coach is a form of incompetence or weakness, but that's the wrong way to think of it because the people who benefit most from coaching are already high performers. These are individuals committed to pushing themselves to the limit and doing everything in their power to gain a competitive advantage. Nobody knows this better than professional athletes.

I had the privilege of interviewing 23-time Olympic Gold medal–winning swimmer Michael Phelps. He is one of the greatest athletes to ever walk the face of the earth, and he talked to me about the power of coaching. Phelps openly admitted he would never have been able to achieve what he did without his coach, Bob Bowman. He credits Bowman with making him hyper-aware of the areas that he needed to improve. There is nobody better than a good coach to help point out blind spots so you can learn what you don't know you don't know.

Whether you're an Olympic swimmer or an everyday investor, I believe you can benefit from the support of a coach. We tend to think of investing as an act done in private or with an advisor. Few consider that investing could be improved with support from a coach or a community. Nobel Laureate Harry Markowitz refers to coaching as a decision-support system, and this is the key trait you want in a financial advisor. Remember, investing is not a get-rich-quick scheme. It's a lifelong journey, so you want a process and a support system for the long haul. When incorporating this strategy, you will work shoulder to shoulder with an advisor/coach who does the following:

- Helps you analyze your existing portfolio to find discrepancies and weak points based on the 20 must-answer questions
- Trains and helps you to develop a deep understanding of academic investing principles, while reinforcing your commitment to those principles
- Helps you to explore and identify how your biases, emotions, and instincts will affect you personally so you can develop a decision-control system that prevents your amygdala from hijacking your investing goals

- Helps you navigate the noise and chaos of the market during both good times and bad
- Helps you construct a portfolio that best suits your personal risk/return preference
- Helps you remain focused on your true purpose for money and life so all of your investing decisions align with that purpose

This is a person who adheres to the academic principles laid out in this book and shares the same investing philosophy as you do. This is a person who would rather lose you as a client than just tell you what you want to hear. And this person has to be a human being—no robo-advisor will do.

What you must do as an investor may seem logical and obvious on paper. However, it's much more difficult to remain disciplined when times get tough and every instinct in your body screams for you to do the opposite of what the research suggests. During those moments, the path of least resistance is incredibly enticing. Your instincts and that voice in your head will try to persuade you to do what makes your brain feel comfortable and safe at the moment. And doing what's comfortable and safe at the moment is not how you achieve long-term investing success. You can have the right asset mix for your desired level of risk and the perfect algorithm to help you rebalance when necessary, but academic investing principles alone will not save you. I've witnessed too many investors and advisors, people who claim to know the science, implode when times get tough. If you don't have the right person in your corner helping to control for natural human behavior, you can panic, pull the trigger, and blow the whole thing up at any given time. That one decision can cost you a lifetime of wealth. Without a coach, a human being who will help you remain dedicated to your commitments, any investing system is bound to fail.

There are two basic types of coaching: one-on-one and a community. The latter is the format on which I built my American Dream Experience workshops. Community is critical, because in a group of 200 to 250 people you aren't only listening to lectures from presenters: you're learning from and seeing yourself in other people in the class. You benefit from their stories, breakthroughs, and insights which can lead to discoveries that you are unlikely to have on your own. By looking at your life through so many different lenses, you can accomplish more personal development when part of a group than you could ever hope to achieve on your own. It's one thing to read the transcript for abject absurdity thought experiment I conducted with Lauren in Chapter 8, but it's something completely different to be there in person and watch it happen in real time. When watching others confront their money demons and dismantle the no-talk rule, you can't help but do the same thing yourself. When you share your purpose for money with others, you help to create it and bring it to life. Nobody thinks of investing in these terms, but when tackling these life strategies together with the help of a community, it becomes a powerful transformational process.

With an advisor/coach who fits these criteria and a community in your corner you will truly be able to align your time, energy, and money with your purpose, so you can live an extraordinary life. As legendary UCLA basketball coach John Wooden said, "A good coach can change a game. A great coach can change a life." However, working with the right coach doesn't let you off the hook or mean that you can take your foot off the gas. In some ways, your work is only beginning, and the road ahead is not easy. But with the right strategies in place, you can better set yourself up for future success.

The Strategies for Living an Extraordinary Life

After being diagnosed with an aortic aneurism, on March 22, 2023, I underwent open-heart surgery to have my ascending aorta repaired. That experience would be one of the greatest challenges of my life. Sometimes, what you believe to be a curse might save you.

Before undergoing surgery, I was perfectly healthy. My condition was hereditary, so there was nothing I could have done to prevent it. I didn't even know about it. The only reason doctors discovered the aneurysm that could have easily killed me was a result of the CAT scan they performed because of my asthma at the time. Without that scan, I might not have undergone the surgery I needed to survive. And it wasn't a simple surgery. The doctors cracked open my chest and dropped my body temperature down to 65 degrees to stop my heart and lungs. They then cut out a piece of the aorta and replaced it with a piece of hose. During the surgery, I was legally dead for 16 minutes—they didn't tell me that part going in.

We are surrounded by medical miracles, and it's because of those miracles that I am alive. The surgery went well, and I have complete heart function today—my aorta is better than new. But that was a scary time and required a long healing process that lasted several months. I couldn't help but emerge from that experience, having learned a series of lessons that have transformed how I see the world.

I've learned that it's critical to have a spouse who loves and supports you through good times and bad. Melissa was by my side every step of the way and helped nurse me back to health. She's an amazing woman, spouse, and warrior. For that, I am forever grateful.

I've learned that time is not infinite. It can seem very long when we're not focused, so we can easily forget about how short life is. That's what makes life so precious and why it's important to cherish every single day. None of us know when our life will come to an end; all we can do is determine how to use the time we have.

I've learned that life can be painful, and that pain is not always fleeting. I tend to think that if I get hurt, I can heal and quickly get over it, but pain can last for months or years. Some people experience long, drawn-out pain and suffering until they die. It's something that we might have to face or help others get through. That reminded me of how important it is to take care of your health, and not take it for granted. To enjoy life while we can, we must prioritize nutrition, exercise, and sleep. I thank God for my blessings and appreciate how lucky I am to be alive. However, there was something else that helped get me through that difficult period.

When recovering, I kept thinking about my purpose. What kept me going was knowing that I could continue to coach and train people to live an extraordinary life by transforming how they invested. I've become more committed to that purpose than ever, and continually search for new ways to help others. I've since learned that the best way to do that is by teaching people how to make an extraordinary life of freedom, fulfilment, and love their natural way of being. This doesn't happen overnight, and it will require hard work and discipline, but there is a process that will get you on the right path. That's what this book is all about. I structured it the way I did because each concept builds on the previous concept to help ensure that change and improvement occur naturally. The entire process can be broken down into these seven actionable strategies.

Ask Powerful Questions

Before you do anything, you must make sure you have a commitment to learning. Even more important, you must be open to learning. That means not just taking "what you know you don't know" and making it "what you know you know." You want to be curious and ask questions so you can learn "what you don't know you don't know." That's what will make all of the difference. Go back and revisit all of the chapter questions. You'll discover that your answers have evolved. And the more you ponder these questions, the more your answers will continue to expand. Remember the questions are really the answers. I have committed my life to these and other inquiries, and they grow deeper and more meaningful over time. Be on the lookout for signs of cognitive dissonance. You can't change your natural human instincts, but now that you know what cognitive dissonance is, and understand how it occurs, you can better identify it. This can prevent you from getting stuck in your ways, doing things because you've always done them a certain way, and failing to identify those things that you believe that simply aren't true. Never stop questioning how the world works.

Develop the Screen of the American Dream

When you change your screen, you change your beliefs, which will change your thoughts, then your actions, and eventually your results. And the only way to get different results in life is to take different actions.

When I was younger, my strategy was "winning at all costs," no matter how hard I had to work. I hoped to overcompensate for the voice in my head that told me I was not smart enough and a coward who was too afraid to fight back. I later replaced it with a new screen—one I fought fiercely to develop over time. My guardian belief and new screen became "Yes, I feel fear and that is only human. And I must diligently work to get results in life. Who doesn't? But my worth does not come from winning. My true worth comes from being a precious child of God." Because of that new screen, I realized that I could fulfill my purpose in life by serving others. I can say with 100% certainty that the success I've achieved personally and professionally is because of the screens I've developed, specifically the screen of the American Dream. I can also say for certain that if I had kept the screens I had when I was younger, I would not have achieved the same success.

The screen of the American Dream is rooted in the freedom and opportunity that is possible because of capitalism and free markets. When seeing the world through this screen, there is no room to behave like a victim or expect something for nothing. You might not realize it, but complaining about your personal circumstances implies entitlement. Whenever you complain about something, it's because you feel entitled to receive or experience the opposite. I still catch myself doing this, and when in that state, I have no power to take action and create the impossible because I have made myself a victim of my circumstances. The trick is recognizing it and understanding that entitlement is the opposite of the screen of the American Dream. When seeing the world through the screen of the American Dream you intuitively understand that you and only you are responsible for your future. It's about working hard to create value for others, and being rewarded for that value you create.

Your screen directly correlates to your results in life, and having the screen of the American Dream will enable you to take actions that someone without that screen won't because they don't believe those results are even possible. So, if you are not where you want to be financially, emotionally, or spiritually, and if you don't have the relationships and level of physical health you desire, the first step is to change your screen. Then, you don't have to create the American Dream. Once you have this screen, the American Dream creates you. You don't have to wake up in the morning and ask yourself how you're going to create the American Dream because that is already how you see the world.

Transform Your Relationship with Money

The single best way to begin transforming your relationship with money is to talk about the no-talk rule. If you don't allow money to create feelings of anger, resentment, and jealousy among your friends and family, you are already ahead of the game and better off than most people. You can take that to the next level by doing battle with your money demons. Don't view money as a scarce resource or a source of survival. Look at money for what it truly is: a phenomenon of language. Money is what you say it is, so make it a powerful tool that you can use to help live an extraordinary life.

Establish Your True Purpose

With the screen of the American Dream in place, you can then turn your attention to identifying and living your purpose. Purpose breeds intention. It's both your guide and your fuel, so all aspects of your life, including investing, begin to line up in the pursuit of that purpose. That's crucial because, when it comes to purpose, you're either moving forward or backward. There is no remaining stagnant. The moment you stop, you get pulled backward, just as you would by the current of a river. When you procrastinate or your actions are not in alignment with your purpose, you send a message to your subconscious that this pursuit is not a priority. If you don't yet know your purpose, help others achieve theirs, and you will eventually discover yours. Discovering your purpose is a non-negotiable part of this process.

You are not your portfolio or any of the material possessions you have accumulated throughout your life. You are not your job. You are your actions and behaviors. You are your purpose. When fulfilling your purpose, you will see the world in ways you didn't before. You will devote your most precious resources—your time and energy—to that pursuit, so the actions you must take to live that purpose will become second nature. Everything in your life will align with your purpose.

Creating an extraordinary life requires you to be a leader. Not a leader who creates followers, but a leader who creates other leaders. Leadership is not something you're born with, but it can be developed, and that occurs with help from your team, or what Napoleon Hill refers to in *Think and Grow Rich* as your mastermind. These are people who align with your purpose. During my interview with Arnold Schwarzenegger, he explained how he never refers to himself as a self-made man because he received so much help from others along the way. However, it's important to understand that it takes time to establish your team, and there will be those who support you only in theory. They seem to be on board when you struggle or face dire circumstances, but once you overcome obstacles to become successful, they often turn resentful and jealous.

Create a Powerful Future View

When you see the world through the screen of the American Dream, and you have established your purpose, you want to use future-based language to paint a clear picture of the world you hope to create. This occurs by making commitments to how you want to live and what you hope to achieve. Your language should always be in alignment with those commitments.

Don't be afraid to dream big. It's perfectly okay if you want to retire early, so you can move to Boca and play golf, but to live an extraordinary life, you must dare to imagine the impossible. That's what Walt Disney did when he came up with the idea for the theme park. This was something nobody had ever done before, and some people thought he was crazy, but he had a vision. That vision was born from his purpose, and that vision changed the world. If you're going to accomplish something extraordinary in life, you must imagine it first.

It's important to understand going in that there will always be those who will doubt and discourage you. You can't let ridicule and criticism prevent you from doing what you want to do. This process has nothing to do with being liked or fitting in. Seeking approval and acceptance is a terrible screen to have because you become more concerned with the opinions of others. That will inevitably lead to you trying to fit in instead of standing out to create your own path, which is essential if you want to live an extraordinary life.

Take Massive Action

At the start of every year, people make commitment after commitment to better themselves—eat better, exercise, strengthen their relationships, become more spiritually connected, and maybe even do all of those things they failed to follow through on the previous year. They usually start the year filled with enthusiasm, only for that enthusiasm to fade before they eventually quit, and nothing changes. That's when some convince themselves they never wanted to do that new thing anyway. Sour grapes. That's what my father used to call it. It's also called being a victim of your own commitments. Either way, it's an excuse, and in the end you will either have excuses or results.

It may seem obvious, but it's worth reiterating because if you don't take action to follow through on your commitments, all of this is meaningless. The investing tools won't produce any results if you don't take action. Your life will not change if you don't take action. You must spend your time, expend your energy, and invest your money to make that future world a reality. In other words, you must practice what you preach. And now that you know what you want, go after it! Eliminate the cancer of casualness and pursue your purpose with a newfound sense of urgency. However, not all action is equal. When pursuing your purpose, you must first learn how to say no to the things you don't want in your life before you can say yes to what you do. If you're a yes person who tries to please everyone, that's the surest way to come up short.

Understand going into this process that you will encounter resistance in the form of fear. One thing I learned as a kid when standing up to bullies was that bravery is not the absence of fear. Bravery is feeling fear and doing the right thing anyway. Much like criticism and ridicule, fear is another potential sign that you're on the right track because it means you're getting out of your comfort zone to do something challenging and new. That's the only way that you can grow, and when you fight through fear, you are rewarded when you get to the other side. It also becomes much easier to do it again.

Through this process of declaring what you will create, making commitments, and following through on those commitments by taking massive action you will become a natural leader. That comes with responsibility and knowing how to harness the power of your word. All great leaders understand the importance of keeping their word by taking actions in line with their commitments. It's what makes your team feel safe and lets them know you have their best interest in mind. That's how you gain their respect, and support. Without your word, you're powerless to create change. Not only will this make you less credible with your team but also you're sending yourself a subconscious message that you can't follow through on your commitments.

This may sound daunting at first, but this is why you can't skip steps. Having established the screen of the American Dream and your purpose, you will more naturally take actions that are in alignment with your beliefs and your vision. That makes this process of taking action much easier and enjoyable. Yes, this work can be fun!

Embrace Academic Investing Principles

This is the final piece of the puzzle that brings everything together. Real dreams require real returns, and when you properly invest, it becomes fuel for your purpose, which can better help you do good in the world and live an extraordinary life.

If there is one secret to investing, it's that there is no secret. Don't let anyone tell you otherwise. It requires hard work, dedication, and discipline. It requires you to remain true to the seven declarations. Gone are the days of speculating and gambling with your money. Discover the answers to the 20 questions so that you understand what's in your portfolio and the level of risk you're taking on. Find an advisor who shares the same commitment to academic investing principles.

Remember that you don't need to do this alone. Work with a coach who has already achieved the type of success you seek who can help you remain disciplined when times get difficult. But to be coachable, you must get rid of the "yeah, buts," so you can listen to your coach, even if you don't understand or agree with that coaching. You need to turn yourself over to the process, so you can exceed your limits and learn *what you don't know you don't know*.

By its very nature, investing can be noble, even beautiful. It's how you can provide money to companies that create breakthroughs in medicine and science that can improve the quality of life for people all over the world. That is a wondrous thing and a far cry from greed and avarice. It's an example of how money and investing can be a form of great good in the world.

****

This isn't the type of book you quickly skim through once to pick up investing tips and life hacks. It's not even a book meant to help you build a better portfolio. It's meant to help you build a better life. This journey is not easy. It's a lifelong pursuit, and there are no shortcuts. The only way to accomplish this is through total immersion.

When you want to learn something and retain that information, you must totally immerse yourself during that learning process.

Immersion is deep mental involvement and focus. Look at the work habits of the best athletes. They don't take days off. Michael Phelps told me how he practiced every single day, including holidays. Kobe Bryant was famous for getting to the gym at 5 a.m. every morning, even the day after the Lakers won the NBA Championship.

I was reminded of how important total immersion was when writing this book. I bought an Airstream trailer and drove it out to the middle of nowhere to isolate myself for days at a time. It was during these isolation periods that I'd completely immerse myself in the material. As I grappled with various problems, they became ingrained in my subconscious. When you reach that stage, your subconscious will work to answer even the most difficult questions. I'd be out cycling and completely in a world of my own when, out of the blue, BAM! I was suddenly hit with the solution to a problem that once seemed so elusive.

Repetition is a crucial part of immersion. That means constantly revisiting these concepts and continually asking yourself the key questions posed in each chapter. Engaging in these thoughts and behaviors over and over is how the brain makes mental maps, strengthens the synapses and connections, and essentially rewires itself. It requires making commitments and then following through on those commitments by taking action.

When you become responsible for your future, something amazing happens. You take back your power, and then you can make almost anything possible. You won't only experience the freedom, fulfillment, and love that comes from living an extraordinary life; you will become the embodiment of it because that is the natural outcome when applying these strategies. The American Dream will become a part of your DNA.

What steps will you take to start living an extraordinary life?

The American Dream Is Eternal

There were many weekends when my father dragged me and my brothers to his office while he put together proposals for his clients. When there, we dutifully did our homework, which he always called "nightwork." However, the weekend of my 15th birthday was special. That Saturday and Sunday belonged to the family alone, and we spent it together. Two weeks earlier, Grandpa George had stormed out of our house, leaving us all stunned and hurt, but our painful feelings had subsided, and our lives had returned to normal.

That Saturday morning was clear and muggy. After the previous night's thunderstorms, only a few clouds remained, and they were quickly evaporating in the morning sun. On the weekends, it was typically my job to mow the grass with our push lawnmower. It was not a big yard, so I could knock it out in about half an hour, but that day I knew the lawn could wait until Sunday. Because it was my birthday, we were headed to Winton Woods for a picnic to celebrate.

My mother had gone to the local IGA, a small chain grocery store, in preparation for the big day. She had stocked up on soda, chips, beer for my dad, candles, and of course, a large chocolate sheet cake, my favorite. My brother Brian had begged her mercilessly for a newly released Elvis Presley record. It was called "Moody Blue," and it was a special edition stamped in blue vinyl. He played it nonstop in the family room, and I could hear him singing along. He had an excellent voice and had memorized the words to all of the king's songs. In good condition, that record is worth more than $1 thousand today. Ours is long gone. It wouldn't have mattered if we had kept it; we wore off the grooves.

Everyone in our family absolutely loved Elvis. We all agreed that he was the best. Is it a coincidence that he helped create a musical genre called *rockabilly*? I don't think so. That title song would reach number one on the charts. We had no idea at the time that record would be his last. When he died less than a year later, my mother cried, and we were all heartbroken. He was only 42, and we simply could not believe he had gone so young.

When someone who seems larger than life dies, it still makes me contemplate my own life and how I am living it. Elvis's money, power, and fame did nothing to protect him from his demons. In fact, the money may have exacerbated the misery he suffered.

Money alone can become a prison just as surely as poverty. There are too many cautionary tales of the rich and famous living lonely and destructive lives that end tragically short. Heavy is the head that wears the crown, or so it seems. Money can become an obligation and drudgery. Wealth, its creation and maintenance, becoming its own kind of weight, pushing down on the creator. And if the money is not made in large sums, well then, the earner might believe that they let everyone down who was counting on them.

My father always coached, "Do what you love, and the money will come." But what if it does come and weaves a curse of its own, creating distance from the people you love—not pulling them closer to you, but pushing them away? It appears to me that living the American Dream and creating an extraordinary life means to somehow escape poverty, but also not to let money become your master. And perhaps escaping poverty itself was the easier of the two challenges.

My father is somewhat of a paradox. I have only seen him cry one time in his life. That happened the day his sister Evelyn died. But while he was tough as nails, he also had an open and vulnerable side. He loved poetry and often read it to me and my brothers. One of my favorites was *If*, by Rudyard Kipling. He loved it so much he memorized it.

> If you can dream—and not make dreams your master;
> If you can think—and not make thoughts your aim;
> If you can meet with Triumph and Disaster
> And treat those two imposters just the same
> If you can bear to hear the truth you have spoken
> Twisted by knaves to make a trap for fools,
> Or watch the things you gave your life to, broken,
> And stoop and build 'em up with worn-out tools.

This was the poem he read to me that day on my 15th birthday. As he shared this excerpt from Kipling's masterpiece, I knew he wasn't reciting it just because it was entertaining. It was because of the lessons he hoped to build into his boys. He was always coaching, training, and developing us as the poem says in its final verse, to become a "man." He wanted us to be strong, so we could stand up for ourselves, but also kind and gentle so we could experience love for family and country.

My parents had rightly suspected, well before they achieved financial success, that it was not the money that made you truly wealthy. It was how you invested your time and energy that gave you a life worth living. It wasn't about the money. It could never be about the money.

In the 1980s, when the real estate market crashed during the savings and loan crisis, there was paralyzing economic fear in the financial markets. The same could be said in 1973 and 1974 when US stock markets dropped in the ballpark of 40% to 50%. My father spent much of his time during those "dark years" reassuring investors, helping them to stay calm and not to panic.

He told me one day, "Mark, there is no need to panic or worry. Hunger can be cured by a can of beans." It was such a good quote that I was sure he had borrowed it from someone else, but I've since searched the internet and can assure you otherwise. That was pure Joe Matson; just as his "nine cents in a phone booth" quote. It was hillbilly wisdom at its greatest—advice I would heed the rest of my life and pass onto you.

That morning of my birthday, after stopping at Kentucky Fried Chicken to grab a king size red and white stripped bucket of chicken with all the fixings—mashed potatoes, baked beans, coleslaw, and biscuits with plenty of butter—we headed to the massive park of more than 2,500 acres. Because it was a beautiful day, the park was crowded, but we found a place by the water and set up our area. We covered the picnic table with the cloth my mother had packed and spread our toys out on the lawn. We had left the fishing poles behind, but we brought three Frisbees (in case we wanted to play Frisbee golf), a football, a badminton set (which took my father and me half an hour to put up), and a set of now outlawed lawn darts. They were later made illegal in 1988, just three years before I would open my own business. We enjoyed throwing the oversized darts high into the air, trying to get them to land in the yellow rings on the ground. Because it had rained the night before, the ground was still soft, and the darts stuck firmly when they hit.

Our mother warned us to "be careful with those things! You could get one stuck in your head." We were sure she was exaggerating but, in my research, I later found they were crafted after a Roman weapon used in military campaigns. If not deadly, they were an emergency room trip waiting to happen. And they were responsible for multiple fatalities, mostly children. Growing up certainly had its own set of dangers and mysteries to be navigated. Luckily, we safely avoided any serious injuries that day.

The picnic table was in the shade and the tan Coleman cooler on the ground next to it. We each grabbed a drink and loaded up our paper plates with food. Because it was my birthday I had first pick, and I went for the dark meat—legs and thighs were my favorite.

As we ate our food, my mother put 15 candles on my cake—that year would be my first in high school. We had the transistor radio (powered by D cells) on the table as we ate. Billy Joel's song "The Stranger" played over the cheap speaker, the piano man, was telling it straight and true about how life can be hard. About how people are not who they always seem to be and how no one ever lets you see fully behind their mask.

The verses of the song got me thinking about how hard life could be, and how George had stormed out of our house two weeks earlier had overwhelmed me. I realized that I had not fully processed it. Maybe there was still a lesson there I needed to learn and catalog for the future. Perhaps I had been too hard on George. I couldn't possibly put myself in his shoes and know what he had dealt with in life. What monsters he had faced. Which ones he had vanquished just to get out of that coal mine. A little gratitude would go a long way. After all, I was where I was in life because I stood on the shoulders of my ancestors. George was stubborn and hardheaded. I wondered how much of that had been passed down to me. Probably a lot.

"Dad," I said, with a mouthful of chicken, "when did you know you were going to get out of the hollers? How old were you?"

"The first time I clearly remember my intuition that I was going to get out was around nine years old. I can't really explain it. I was walking home from school. I was upset I didn't have shoes for gym, and that the bigger kids had made fun of me. I was fighting back tears, and I just knew. There wasn't even any doubt about it. Then, the sadness went away, and I was just happy."

"Mom, what about you?" I asked.

"I always made straight As in school," she said. "I'm not bragging. I just did. When I went to high school, I could visualize getting out of there. The richest part of Charleston, West Virginia, is South Hills, but I didn't even see myself there. I just knew I had to get completely out. Your father and I didn't even need to talk about the decision; it was almost like telepathy. We just knew, and it was a strong feeling."

Although many people living there had just accepted their fate, it hit me that fate can work in two ways. For if you believe hard enough, you can change your fate to live a new and improved life. Like my parents, first you must clearly see it and come to believe it. Then, you must be willing to fight for it.

In an odd way, my parents' story reminded me of the Grimms' fairy tale about Hansel and Gretel. Like the two small children in the story, they had largely been abandoned by their parents. Hansel and Gretel were taken out into the wilderness to die by the people who should have fought to protect them from the harsh evil of the world. But the children were industrious and left breadcrumbs to find their way back home. When they were taken out into the woods a second time, they found a house made of cakes and candy. An evil witch inhabited the tempting home and began to fatten up the children, so she could later consume them. When Gretel learned of the witch's sinister plan, she bravely threw the hag into her own oven and killed her.

Symbolically, my parents did not just escape the evil and the destitute poverty around them; they fought that evil and destroyed it by creating a new reality for our family. The witch had been defeated, and the children had returned home with her treasure.

But in my family's story, the treasures were not made of gold, silver, or jewels, but of love, purpose, and self-expression. I saw our family saga as a clash between good and evil. In our story, good carried the day.

While I was lost in thought, Matt suddenly blurted out, "Why do you always have to talk about this serious stuff? Come on, this is a birthday party. Make a wish already and let's eat some cake."

Brian said, "News flash! Special report! Mark is trying to figure out the nature of the universe—again. Hey, big bro, leave the world alone long enough to have some fun because—Live from New York It's Saturday Night."

We all smiled at that as my mother lit the candles on the cake, and they all sang a rousing rendition of *Happy Birthday*, ending with, "you smell like a monkey, and look like one, too!" So-called magic birthday candles that relight themselves after you blow them out cost five times as much as the regular ones. On this occasion, Mom had splurged on the magic kind, and the entire family laughed as they watched me try to blow them out. After several unsuccessful attempts, I rolled my eyes and then gave it one last massive effort that finally extinguished them.

We stayed for several hours tossing the football, playing Frisbee golf, and trying to amuse ourselves with badminton, which we were all terrible at. The game wasn't graceful, but it was good for a ton of laughs. It was a short ride home, but we were all thoroughly tired. As we pulled into the driveway, I saw the aluminum eagle (Figure 12.7) my father had bought at Central Hardware anchored securely over our garage door. It was a talisman, the embodiment of freedom intended to ward off the violence of victimization and entitlement. It was there to remind us that our future was ours to create, and that the American Dream is real.

The eagle enabled us to see that we had love and connection, the kind that makes its own magic to ward off evil. The kind that enabled my brothers and me to know we would never be abandoned in the wilderness to fend for ourselves. Our love for each other was our treasure. It was one that no evil force could take away.

Figure 12.7   The Eagle Hanging over the Entry of My Childhood Home

Notes

1. Burton G. Malkiel and Charles D. Ellis, *The Elements of Investing: Easy Lessons for Every Investor* (John Wiley & Sons, 2010), pp. 49–50.
2. Ibid., p. 49.
3. Ibid, p. xiv.
4. Guido Carim Junior, Chris Campbell, Elvira Marques, Nnenna Ike, and Tim Ryley, "Almost Half the Men Surveyed Think They Could Land a Passenger Plane. Experts Disagree." CNN Travel, December 6, 2023. https://www.cnn.com/travel/how-easy-is-it-to-land-a-passenger-plane/index.html.

See Appendix II for additional information regarding data in this chapter.

Conclusion: It Won't Get Any Easier

With the help of my family and team, we have successfully built a company with more than 60 employees, 500 independent advisors, and $10.7 billion under management as of May 2024. We have served more than 35,000 families and investors in the United States, Puerto Rico, and Canada. And what happened to the old broker-dealer I worked for? It was acquired by another mega company and struggled to sell its big plot of land where its home office sat empty for years. During my career, there were ups and downs. I did many things right. However, there was one thing that I was wrong about in the beginning, and wrong in a massive way.

I believed that once advisors saw the academic research, they would never speculate or gamble with their clients' money again. I imagined a grand investing enlightenment that would end stock picking, market timing, and track record investing. I figured that, with my business model, I could last maybe 10 to 15 years tops. That's when there would be a brand-new utopia in the world of investing where everyone could see the errors of their previous ways. It just didn't make sense for anyone *not* to invest according to empirically tested academic research. But, oh boy, was I wrong. Was I just a naive 27-year-old, or did I miss something bigger? Maybe it was a little bit of both.

First, I was overly optimistic about the power of knowledge. I later discovered that simply knowing how to do the right thing doesn't ensure taking the right actions. Knowledge doesn't equal action. I also couldn't foresee 24/7 news channels. In the 1980s, the news aired at 6:00 p.m. and 11:00 p.m. It was a half-hour of straight news. Now, you have entire networks dedicated entirely to the markets and providing investment advice. And what does that advice involve? Trying to predict the future of political and economic events in the futile attempts to beat the markets. It doesn't matter which station you tune into; it's their job to prey on your fears and make you miserable with the single goal of getting you stimulated. That's it. These stations don't exist to make you happy, give you sound financial advice, or try to make your life better. They exist to scare the heck out of you because that's what makes them money. Fear equals more eyeballs and that equals higher advertising fees from the program's sponsors. This makes the academic investing principles extremely hard to follow.

I also couldn't foresee the proliferation of all the toxic investment vehicles people use today. I'm talking about poorly designed index funds, ETFs, cryptocurrency, NFTs, ESG investing, and the never-ending list of vehicles being pushed on the investing public. What you need to understand is that the investing world is on a mission to offer more exotic and toxic products that can destroy your wealth and make gambling seem attractive and easy.

And 35 years ago, the smartphone was nowhere on the radar. I didn't imagine it or its impact on the world. Who did? It fundamentally changed how we live, work, and invest. There are more ways for you to lose your money than ever. Today, anybody can open their Robinhood app and start gambling with their money, just like they would on their DraftKings app. Technology exacerbates this problem. It's so addictive and destructive because now people can do all those nasty things with their money that they shouldn't be doing in the first place. Any time they get the urge to trade, they can use this small device they carry around in their pockets. The line between prudent investing and destructive gambling has been increasingly blurred.

I think the world would be a far better place if people would admit how they have been wrong. It can help us look in new directions and inform us about how the world really works. I constantly remind myself to be curious about the world and be willing to see how I have been wrong. Some of my biggest breakthroughs in life have come from having this mindset. Not only was I wrong about people using academic research but also the science of investing has been largely hijacked and turned into slick marketing jargon, but it's not effectively used. And now, I believe it's only going to get worse. At this point, I don't ever expect it to stop, which is why it's more important than ever to learn these strategies to create an extraordinary life. When I talk about an extraordinary life, I'm not talking about class, status, or money. I'm talking about living a life with freedom, fulfillment, and love. I'm talking about experiencing the American Dream, and that's something we all deserve.

Epilogue

Arthur B. Laffer

In the book you have just finished reading, you have been witness to a story of hope and determination leading to fulfillment. The man I met, Mark Matson, years and years ago at a financial conference in Chicago, had overcome the odds and was on his way to astounding success. He founded Matson Money and discovered the age-old recipe of doing well by doing good. He has provided all sorts of people from any number of professions with sound advice and tangible options to preserve and augment their hard-won savings. For most people, investing is not easy or natural, and with only a few missteps, all could be lost. Preserving wealth is a treacherous venture at best, and those with wealth need to be very careful not to make a rookie mistake. By using asset diversification and academic research, disciplined investing could materially reduce the risk of failure and enhance the prospects for financial security throughout one's life. In cliché terms, it takes a long career of hard work and focus to be an overnight success. The keys are to stay the course, don't panic, stay disciplined, be positive and optimistic, and never ever ever give up.

The Matson story is an all-American story, resonant with that of the United States itself, on the world stage. As history reveals, most attempts at new ventures fail. But in many cases, failure, hard-knocks experience, and setbacks have only caused genuinely optimistic happy warriors to redouble their efforts. Since the mid-1960s, I've been part of numerous episodes of trials and tribulations as a fly on the wall of politicians, entrepreneurs, innovators, pathbreakers and illuminati. The greatest success I've been intimately involved with was the political career of President Ronald Reagan.

299

And believe you me, there were numerous ups and downs from the day he entered the governor's race in California's 1966 primary election to his farewell address as president of the United States in 1989.

Reagan's own pursuit of the American Dream began with his rise from his humble beginnings in Dixon, Illinois, to stardom as a Hollywood actor. We also all know of Reagan's final triumph as president of the United States, propelling America to once again be the economic, military, and democratic free market leader of the third rock from the sun, instilling the ideals of the American Dream for millions of people throughout the world. But were you aware that Reagan was also a devoted union member who became head of the two actors' unions, AFTRA and SAG? And, as president of the unions, he was the first to call for a bitter nationwide strike against the studios?

Were you also aware that the policies he promulgated as California's governor included large tax rate increases in the already high personal income tax, capital gains tax, and corporate tax? He also was the biggest social spender back then. He chaired the Equal Rights Amendment, eliminated California's anti-abortion statutes and bashed President Kennedy's tax cuts that took effect from the early 1960s through 1966. How could that man turn out to be the greatest US president ever? The answer is as simple as it is profound: he learned. He learned from his mistakes and his successes. His entire career he spent learning and changing when the facts demanded. He was special. He wasn't born great; he earned his greatness. Reagan followed a similar path as what has been written by Mark in this book: he believed in the American Dream and continuously worked hard so that not only his dream could be realized but also so others throughout the world could also pursue and realize theirs.

By the time he finished his second term as president, President Reagan had cut the highest federal personal income tax rate from 70% to 28%, inspired the cut in the capital gains tax rate from almost 50% to 28%/20%, reduced the corporate tax rate from 46% to 34%, and eliminated massive amounts of deadweight tax loopholes.

He, more than any president ever, came the closest to achieving the ideal tax policy of the lowest possible tax rate on the broadest possible tax base: the North Star of tax policy.

His economic policies in conjunction with his focus on foreign policy reduced inflation from double digits to low single digits, increased employment by 20+ million jobs, defeated the Soviet Union's military threat (without a war), ended major riots in the US, oversaw a huge reduction in union membership and labor strikes, and put in motion a worldwide movement of pro-growth democratic economic capitalist tax cuts and other pro-growth policies.Today, our nation appears to be irretrievably split. But it was also seemingly irretrievably split in the mid- to late 1970s. Watergate et al. was not a love fest. Everyone was screaming, accusing, and behaving badly to their fellow Americans—much like today. Ronald Reagan rose to the occasion and brought all of us together on a message of hope and peace through strength. His mission to unite the country reminds of Mark's writings about bringing families together by helping them to talk about money and providing purpose to their lives so that they can pursue their American Dreams in a united fashion.

President Reagan was re-elected in 1984, carrying 49 out of 50 states. In fact, as a member of the Executive Committee of the Reagan/Bush Finance Committee, I was keenly aware that President Reagan, early on, pulled his financial campaign commitments out of Minnesota to allow his opponent Walter Mondale (who was from Minnesota) to carry his home state. President Reagan was the very definition of a uniter, of a gentleman, of a patriot.

In 1986, Reagan's signature piece of legislation, the 1986 Tax Act, passed the Senate in its final vote 97–3. Everyone was on Team Reagan. One nation under God with liberty and justice for all.

Reagan let America become great again, opening up after a dim period the possibility of wealth accumulation and the handing down of wealth productively from one generation to the next. Mark Matson and his father Joe were wise enough to see this opportunity and offer their sage direction and advice to those prospering in the great Reagan-inspired American economic resurgence.

Reagan was able to bring freedom and fulfillment to millions of people not only in America but throughout the world and opened up the possibility of others being able to realize the ideals of the American Dream.

I just love these stories and am honored to have been a part of so many of them.

Appendix I: My Vision Statement

Conceptual Framework for New Space in the Big AZ

The space will be like no other place on the planet. It will be far beyond the ability for anyone to imagine before they experience it. The building will affect the person experiencing it on an emotional level. It will take people on a journey that they can experience in a personal, not generic, way. The building itself will be an experience. There will be no front stage and backstage—it is all front stage. The space will also make it self-evident as to the way that guests flow through the space. Think of the space as you would a ride at Disney. It tells a story, and the guests become part of that story.

Even if a guest tours the facility alone, they will take away the complete story of Matson Money: how they save investors and make the world a better place. The space will draw them in and make them want to stay and explore—it will be compelling. It will be the most amazing building they have ever entered. In short, the building doesn't just become part of the awesome engaging experience—it is an awesome engaging experience in its own right.

The space is a repository of the Matson Money values. This will be explicitly demonstrated but also be part of the very fiber of the space. It will scream—free markets, entrepreneurism, integrity, education, and inquiry. It will also demonstrate the dangers that face investors—Wall Street bullies, the Evil Empire, and the ever-present danger of technology as a destructive force embodied in the robo-advisor. We will also show guests that technology can also be used for good, if harnessed properly. They will feel that with our coaching, they will be confident in their ability to become technology's master. The advisor and the investor will step into this new reality and become the heroes of their own story. It will be clear how, through saving and empowering themselves and the people they love and care about, they will make the world a better place.

They will feel the power of the community that built this space and understand that they can share in it—they will be part of the community. It will become part of their identity. They will harness the power in their own lives. They will also feel at home and know that this space was created for them. They will feel inspired but also safe and supported. They will feel like they are entering another world. One where the little guy can win—good over evil. They will know that we truly care about them. The space will be fun and convey a sense of adventure.

My true purpose in life is love. They will feel loved. They will also take hope in the reality that someone has created this! "If this can be imagined and built, anything is possible!" This space will empower its guests to take massive action. The space will lead to a feeling of hope and optimism about the future. It will help offset the pessimism and negativity created by the media and used by doom-and-gloom market prognosticators. It will convey the benefits of capitalism and show that the right technology is creating miracles every day. The building should put the current "life reality" into context to how things were just 500 years ago—the average person making $30k lives better than the monarchs of England. It will create and spread a deep feeling of gratitude for all we have as Americans.

The space will give a perspective of time—the past, present, and future. It will give perspective to what has passed, what is, and what could be. It will celebrate innovation and the guys-in-the-garage effect. It will have the feel of a place that creates the foundation for a global movement—a headquarters of such, or perhaps a war room. It will create a sense of action and energy.

When advisors come to training, they will be inspired to become true entrepreneurs and have a deep desire to bring their clients to share in their insights. This will be a place of community. The ultimate power of the place will be vested in its ability to create a "tribe" of people who are transformed and want to go back to their local communities and help save their families, friends, and loved ones from investing devastation. The space will demonstrate the power of the community and help the investor see why they want to be a part of it. It will also show the strategic relationship that we have with the advisors and why the coach is paramount to their success as an investor. It will help them understand what we do here and what the coach does and why they need both of us.

The experience will also humanize what is often a cold process. We need to let them know that we understand that this is not just about portfolios but about real people, real dreams. It will also demonstrate that we have a long heritage and real returns—not fake or hypothetical returns that most advisors show to confuse investors. Along those lines, we will show the total growth of assets and highlight all of the crashes and tragedies that have happened during that time. The building will convey the feeling of stability and strength and that we know how to "weather the storm."

History will be a big part of the space and provide context and direction—history of markets, history of Matson Money, and history of wealth creation. We will demonstrate our commitment to the Academic Board. It will be easy to purchase books, videos, and clothing so that they can fondly remember their time with us.

They will also see and feel that we have a commitment to philanthropy and giving back. Our ethics will be demonstrated—ethics books should be readily available. It will be obvious that we are different from any other money management company. They will see this as an experience and not some place to park money. We will convey the value of both money management education and the value of the life coaching and help them to distinguish between the two.

Freedom and patriotism will be another key theme. It will be evident that we do believe that America is exceptional, and that we value the men and women who have fought to keep us all free and give us this amazing freedom. We will convey that we have a powerful vison of the future for America and that we believe in the American Dream.

The floor will change texture as you move from one area to the other to indicate that the topic and next part of the story is being told.

The 20 must-answer questions will be a main theme. The idea that we can help demystify investing—but the investor must be willing and able to work for it. We can work together, but we will not do it for them. We, and only we, will teach them the right questions and guide them through the process, so they can have a successful investing experience and recapture their personal American Dream. Training them to see the market forces and how to harness them and stay disciplined over a lifetime.

The space will largely be interactive—the more they can touch and feel and connect with the better. We will have a phone app that they can use to help guide them on "the tour." Helping millennials and celebrating them and empowering them to grow and be successful is paramount. Give them the handheld technology they crave but also the human contact and direction they are lacking in their life. Promote the symposium throughout. Help them to see how amazing the experience is and why they want to be part of it. Art will also be part of the space to convey credibility and lift people up.

Appendix II: Disclosures

T his content is not to be considered investment advice and is not to be relied on as the basis for entering into any transaction or advisory relationship or making any investment decision.

In this book, I share with you my opinions and recollections. That means the ideas I present throughout this book are just one view. Others have different views, and I leave it to you to judge and decide which approach will best help you meet your goals. Don't take me at my word; I invite you to question, research, test, and challenge these ideas.

Also, you should remember that investing comes with risk, and every investor should think carefully about the risk they are prepared to take. Every investor should also know that past performance cannot tell you what future results will be. Even ideas and strategies that performed well previously have risk and can perform very differently in the future.

In this book, I have portrayed conversations involving others. I have put some of these in quotes to indicate that another person is speaking, but these reflect my recollection, not transcripts or recordings made at the time, and might differ from the recollection of others.

****

The individuals who have been kind enough to support this book by writing a foreword, epilogue, or other statement for the cover are not clients of my firm, Matson Money, Inc. None were paid or otherwise compensated for their supporting words. However, some of these individuals have a relationship with my firm. For example, Mr. Laffer sits on our Academic Advisory Board, and Mr. Lowe has previously appeared at Matson Money events. Others are associated with firms that provide investment products and services used by Matson Money. In some cases, there may have been compensation in connection with the relationship or prior events or the commercial relationship between Matson Money and other firms. As a result, and because the individuals may receive some intangible benefits from being associated with this book (e.g., heightened brand awareness or reputational enhancement), there may have been incentives for their support.

****

This book contains data, graphs, charts, text, or other material reflecting the performance of a security, an index, an investment vehicle, or other instrument over time ("Performance Material"). Past performance, including any performance reflected in Performance Material, is not an indication of future results.

All investments involve the risk of loss, including the loss of principal. These risks may not always be mitigated through long-term investing or diversification. No investment strategy (including asset allocation and diversification strategies) can ensure peace of mind, guarantee profit, or protect against loss.

PAST PERFORMANCE IS NO GUARANTEE OF FUTURE RESULTS.

****

Matson Money, Inc. ("Matson") is a federally registered investment advisor with the Securities Exchange Commission (SEC) and has been in business since 1991. In Canada, Matson is registered as a portfolio manager in Ontario, Alberta, and British Columbia. Registration with the SEC and the Canadian securities regulatory authorities does not imply their approval or endorsement of any services provided by Matson. *For more information, please see the Matson Money Form ADV Part 2A.*

****

Certain information contained in this content may be based on, or derived from, published and unpublished information provided by independent third-party sources and are believed to be accurate and reliable. However, the accuracy of such information cannot be guaranteed. No representation or warranty, express or implied, is made in respect thereof, and neither the author nor his firm takes responsibility for any errors or omissions that might be contained herein or accepts any liability whatsoever for any loss arising from any use of or reliance on this information.

The statements contained in this content that are not historical facts are forward-looking statements, which are based on current expectations and estimates about particular markets. These statements are not guarantees of future performance and involve certain risks, uncertainties, and assumptions that are difficult to predict. Therefore, actual outcomes and returns may differ materially from what is expressed in such forward-looking statements.

****

Academic principles are discussed throughout the book and in detail in Chapter 10. The investing principles primarily discussed are based on the following research.

Factor Model

Fama, Eugene F., and Kenneth R. French. "The Cross-Section of Expected Stock Returns." *Journal of Finance* 47 (June 1992).

Fama, Eugene F., and Kenneth R. French, "Common Risk Factors in the Returns on Stocks and Bonds." *Journal of Financial Economics* 33, no. 1 (February 1993).

Fama, Eugene F., and Kenneth R. French, "Profitability, Investment and Average Returns." *Journal of Financial Economics* 82, no. 3 (December 2006).

Fama, Eugene F., and Kenneth R. French, "A Five-Factor Asset Pricing Model." *Journal of Financial Economics* 116, no. 1 (April 2015).

Efficient Market Hypothesis

Fama, Eugene F. "Random Walks in Stock Market Prices." *Financial Analysts Journal* (September/October 1965).

Modern Portfolio Theory

Markowitz, Harry. *Portfolio Selection: Efficient Diversification of Investments* (New York: Wiley, 1959).

****

The Nobel Memorial Prize in Economic Sciences, commonly referred to as the Nobel Prize in Economics, is an award for outstanding contributions to the field of economics, and generally regarded as the most prestigious award for that field.

Markowitz, Harry. "Portfolio Selection." *Journal of Finance* 7, no. 1 (1952).

Harry Max Markowitz is an American economist and a recipient of the 1989 John von Neumann Theory Prize and the 1990 Nobel Memorial Prize in Economic Sciences. Markowitz is a professor of finance at the Rady School of Management at the University of California, San Diego.

The efficient market hypothesis was first explained by Dr. Eugene Fama in his 1965 doctoral thesis.

Fama, Eugene F. "Random Walks in Stock Market Prices." *Financial Analysts Journal* (September/October 1965).

Eugene F. Fama, 2013 Nobel laureate in Economic Sciences, is widely recognized as the "father of modern finance." His research is well known in both the academic and investment communities. He is strongly identified with research on markets, particularly the efficient market hypothesis.

Historical Performance of Indices

Index data appears in Chapters 5, 6, 7, 8, 9, 10, and 11.

This content includes historical performance information from various global stock market indices. Index information is included in this content solely for educational purposes to help demonstrate academic investment principles being described (e.g., the potential benefits historically associated with diversification of asset classes, the equity risk, small risk, and value risk premiums, etc.) or demonstrating historical events. Comparison to any specific investment or fund is not intended, and the data does not represent or suggest results Matson Money would or may have achieved when managing client portfolios. Investors cannot invest in a market index directly, and the performance of an index does not represent any actual transactions. The performance of an index does not include the deduction of various fees and expenses, or the impact of taxes, each of which would lower returns. Index information assumes the reinvestment of dividends and capital gains.

In some cases, index data is translated into dollar figures. Those dollar figures are derived simply by applying the performance of the index over time to a sample dollar amount. Such dollar figures in the text and charts are intended as a pedagogical method solely to give tangible meaning to the index data being presented. Similarly, where the book makes reference to having "invested" at a certain time in an index, this is not meant literally, but rather as a tool to understand how the asset class reflected in the index data has performed over time. Readers should understand that the goal is NOT to suggest that these dollar figures reflect performance that could be achieved by a portfolio.

PAST PERFORMANCE IS NO GUARANTEE OF FUTURE RESULTS.

Index information is derived from returns software created by Dimensional Fund Advisors LP (DFA). DFA is a registered investment advisor that, among other things, specializes in and sells statistical market research and mutual fund management. DFA obtains some of its market data from the Center for Research & Security Pricing (CRSP), part of the University of Chicago's Booth School of Business (Chicago Booth). In addition, some data discussed in the book might also be obtained from Morningstar Direct.

Data prior to the launch date of the index is created by calculating how the index might have performed over that time period had the index existed.

There are limitations and risks associated with the use of the index information as described here. Investment strategies should not be based solely on the use of such information. Actual investment results will be different from the index results shown.

Appears in Chapter 11 and Appendix I

Academic Advisory Board members for Matson Money generally receive compensation from Matson Money for their services, which include, but are not limited to, independent leadership consulting, coauthoring white papers, and speaking at Matson Money conferences. Advisory Board members may also provide insight to Matson Money on portfolio construction, asset allocation, quantitative analysis, investor behavior, and other areas of expertise, as needed. **Certain Advisory Board members are employed by or otherwise affiliated with third-party advisory firms that offer funds in which Matson Money client accounts are invested.**

Applicable to Chapter 1, 5, 6, and 7

Where data is shown representing company stock performance, the return data shown does not include the deduction of fees or expenses associated with trading, which would lower returns.

Applicable to Chapter 5, 6, 7, 8, 9, 10, and 12

Where data is shown representing mutual fund performance, other expenses, such as account-level fees or expenses associated with accessing a fund, if any, are not shown and would lower returns.

Chapter 1 Figure 1.6: Boller, Steve. "Spaced Learning and Repetition: How They Work and Why." Bottom Line Performance (October 16, 2012). www.bottomlineperformance.com/spaced-learning-and-repetition-why-they-work/.

****

Chapter 5 and Figure 5.4 Additional information regarding "The 21 Best Stocks to Buy for 2021." *Fortune Magazine* (December 2020/January 2021). https://fortune.com/2020/11/20/best-stocks-to-buy-for-2021-airlines-health-care-green-energy-banks-consumer-international/.Average return of 21 stock recommendations for 1/1/2021–12/31/2021. Performance data for stocks collected from Morningstar Direct.Stocks: American Tower Corp, Atlantica Sustainable Infrastructure PLC, Bank of America Corp, Burlington Stores Inc, Canadian National Railway Co., Cellnex Telecom SA, Clearway Energy Inc Class C, Crown Castle International Corp, Hexcel Corp, JetBlue Airways Corp., Live Nation Entertainment Inc, NextEra Energy Inc., Southwest Airlines Co., Stryker Corp., Taiwan Semiconductor Manufacturing Co. Ltd. ADR, Teleflex Inc., TransDigm Group Inc., Unilever PLC ADR, UnitedHealth Group Inc., VF Corp., and Wells Fargo & Co.

****

Chapter 5 and Figure 5.5 Additional information regarding "11 Steady-Rising Stocks to Own for 2022. *Fortune Magazine* (December 2021/January 2022). https://fortune.com/2021/12/01/stocks-to-own-2022-amzn-msft-pypl-crm-pep-jnj-cmcsa-shop-tsm/.Average return of 11 stocks from 1/1/2022–12/31/2022. Performance data for stocks collected from Morningstar Direct.Stocks: Amazon.com Inc, Comcast Corp Class A, Johnson & Johnson, Microsoft Corp, Nestle SA ADR, PayPal Holdings Inc, PepsiCo Inc., Salesforce Inc., Shopify Inc Registered Shs -A- Subord Vtg, Taiwan Semiconductor Manufacturing Co. Ltd. ADR, and Tencent Holdings. Ltd. ADR.

Chapter 7 and Figure 7.1 Data from CRSP and Compustat. Companies are sorted every January by beginning of month market capitalization to identify first-time entrants into the 10 largest stocks. Market defined as Fama/French US Total Market Research Index. The Fama/French indices represent academic concepts that may be used in portfolio construction and are not available for direct investment or for use as a benchmark.

DALBAR References (Chapter 7) Average stock investor, average bond investor, and average asset allocation investor performance results are based on a DALBAR study, "Quantitative Analysis of Investor Behavior (QAIB), 2024." DALBAR is an independent financial research firm. Using monthly fund data supplied by the Investment Company Institute, QAIB calculates investor returns as the change in assets after excluding sales, redemptions, and exchanges. This method of calculation captures realized and unrealized capital gains, dividends, interest, trading costs, sales charges, fees, expenses, and any other costs. After calculating investor returns in dollar terms, two percentages are calculated for the period examined: total investor return rate and annualized investor return rate. Total return rate is determined by calculating the investor return dollars as a percentage of the net of the sales, redemptions, and exchanges for the period.

Chapter 8 Some mutual fund performance information is from the CRSP Survivor-Bias-Free US Mutual Fund Database, which is the only complete database of both active and inactive mutual funds and serves as a foundation for research and benchmarking mutual funds. For more database information, see https://www.crsp.org/products/research-products/crsp-survivor-bias-free-us-mutual-funds.

Chapter 10 Additional information regarding the track record of active fund managers outperforming a benchmark: S&P Global, "SPIVA® U.S. Scorecard Year End 2023." https://www.spglobal.com/spdji/en/documents/spiva/spiva-us-year-end-2023.pdf.

Chapter 11 Performance information discussed in Chapter 11 regarding specific country performance is based on MSCI country indices (net dividends) for each country listed. This does not include Israel, which MSCI classified as an emerging market prior to May 2010. For emerging market individual country information, the MSCI country indices (gross dividends) for each country listed is used. The data does not include Greece, which MSCI classified as a developed market prior to November 2013. Additional countries are excluded due to data availability or due to downgrades by MSCI from emerging to frontier market.

****

Fixed income default information sources: S&P Global Fixed Income Research and S&P Global Market Intelligence's CreditPro®. S&P Global Ratings "2022 Annual Global Corporate Default and Rating Transition Study." https://www.spglobal.com/ratings/en/research/articles/230425-default-transition-and-recovery-2022-annual-global-corporate-default-and-rating-transition-study-12702145.

Standard & Poor's Global Fixed Income Research and Standard & Poor's CreditPro®. http://www.standardandpoors.com/ratings/gfir/en/us.

Chapter 11 and Figure 11.2 An efficient frontier as written about in Markowitz's paper is to show the benefits of diversification into several asset classes with low or negative correlation among the selected asset classes. This chart does not reflect actual performance of any managed portfolio and no representation is made that your portfolio would achieve similar results. The Markowitz efficient frontier graph plots annualized compound return and standard deviation of a portfolio. It shows how investors can either increase potential returns for the same level of volatility or decrease volatility for the same potential rate of return.

Chapter 11 and Figures 11.3, 11.4, and 11.5 R is a correlation coefficient that measures the strength of the relationship between two variables, as well as the direction on a scatterplot. The value of r is always between a negative one and a positive one (-1 and a $+1$).

Figure 11.4 is based on the MSCI EAFE Index and the MSCI Country Specific. Data is shown for countries to be statistically significant. Other countries might be available.

Chapter 12 Information regarding GIPS-compliant firms obtained from multiple data sources: the number of state registered investment advisor firms is obtained from "NASAA Investment Adviser Section 2023 Annual Report"; the number of SEC-registered investment advisor firms is obtained from https://www.sec.gov/help/foiadocsinvafoia, and GIPS compliant firm data is obtained from https://compliancetracking.cfainstitute.org/gips-firm-list.

Appendix III: Index Descriptions

This content includes historical performance information from various global stock market indices. Index information is included in this content solely for educational purposes to help demonstrate academic investment principles being described (e.g., the potential benefits historically associated with diversification of asset classes, the equity risk, small risk and value risk premiums, etc.) or demonstrating historical events. Comparison to any specific investment or fund is not intended, and the data does not represent or suggest results Matson Money would or may have achieved when managing client portfolios. Investors cannot invest in a market index directly, and the performance of an index does not represent any actual transactions. The performance of an index does not include the deduction of various fees and expenses or the impact of taxes, each of which would lower returns. Index information assumes the reinvestment of dividends and capital gains.

- S&P 500 Index

  The S&P 500® Index is widely recognized as representative of the equity market in general. The S&P 500 Index is an unmanaged, capitalization-weighted index designed to measure the performance of the broad US economy through changes in the aggregate market value of 500 stocks representing all major industries.
- Nasdaq Composite Index

  The Nasdaq Composite Index is the market capitalization-weighted index of approximately 3,000 common equities listed on the Nasdaq stock exchange. The types of securities

---

Index data appears in Chapters 5, 6, 7, 8, 9, 10, and 11.

in the index include American depositary receipts, common stocks, real estate investment trusts (REITs), and tracking stocks, as well as limited partnership interests. The index includes all Nasdaq-listed stocks that are not derivatives, preferred shares, funds, exchange-traded funds (ETFs), or debenture securities.

The market capitalization–weighted methodology is used to calculate composite index value. The value of the index is calculated thus. The total index value is equal to the share weight of each of the component securities multiplied by the last known price of the concerned security. The final value obtained is thereafter adjusted by a divisor that scales the value for the purpose of convenience in reporting. The composite index is continuously calculated throughout a trading day and reported once in a second. However, the final confirmed value is reported at 4.16 p.m. on a trading day.

- CRSP Decile Indexes

    CRSP Cap-Based Portfolio Index data are a monthly series based on portfolios that are rebalanced quarterly. Matson uses these indexes to represent specific segments of the domestic equity market. The Center for Research in Security Prices (CRSP) calculates indices for five groups of US stock markets (NYSE, AMEX and NASDAQ separately, NYSE/AMEX combined, and NYSE/AMEX/NASDAQ combined) in which all securities other than ADRs are ranked by their market cap and then divided into 10 deciles with an equal number of securities in each decile. Starting with the NYSE, CRSP first sorts all stocks on the NYSE by market cap and breaks the universe into 10 equal groups, called *deciles*, by number of names. Decile 1 represents the largest stocks on the NYSE and decile 10 represents the smallest NYSE stocks. CRSP then includes all equivalently sized AMEX and NASDAQ stocks into the NYSE size decile in which they fit by market cap.

    i. CRSP 1–10 Index: Representing the entire market cap of the NYSE and other exchange equivalents.

ii. CRSP 1–5 Index: The largest half of NYSE stocks by name and all equivalents from other exchanges, covering large cap through mid-cap stocks.

iii. CRSP 6–10 Index: The smallest half of NYSE stocks by name and all equivalents from other exchanges, sometimes referred to as *small cap* stocks. The CRSP 6–10 Index is similar in size to the Russell 2000 Index.

iv. CRSP 9–10 Index: The smallest fifth of NYSE stocks by name and all equivalents from other exchanges, sometimes referred to as *micro-cap* stocks.

The Dimensional Indices have been retrospectively calculated by Dimensional Fund Advisors LP and did not exist prior to their index inception dates. Methodology used for computing profitability premiums: Dimensional controls for relative price (BtM) and size (market cap) when computing the annualized compound returns for high and low profitability stocks in US and non-US developed markets and controls only for relative price in emerging markets. Profitability is measured as operating income before depreciation and amortization minus interest expense scaled by book. Dimensional index data is compiled by Dimensional from CRSP, Compustat, and Bloomberg.

• Dimensional US Large Cap Value Index

The Dimensional US Large Cap Value Index is compiled by Dimensional from CRSP and Compustat data. This index targets securities of US companies traded on the NYSE, NYSE MKT (formerly AMEX), and Nasdaq Global Market with market capitalizations above the 1,000th-largest company whose relative price is in the bottom 30% of the Dimensional US Large Cap Index after the exclusion of utilities, companies lacking financial data, and companies with negative relative price.

The index emphasizes securities with higher profitability, lower relative price, and lower market capitalization. Profitability is measured as operating income before depreciation and amortization minus interest expense scaled by book. Exclusions: non-US companies, REITs, UITs, and investment companies. The index has been retroactively calculated by Dimensional and did not exist prior to March 2007. The calculation methodology for the Dimensional US Large Cap Value Index was amended in January 2014 to include direct profitability as a factor in selecting securities for inclusion in the index. Profitability is measured as operating income before depreciation and amortization minus interest expense scaled by book. Prior to January 1975: Targets securities of US companies traded on the NYSE, NYSE MKT (formerly AMEX), and Nasdaq Global Market with market capitalizations above the 1,000th-largest company whose relative price is in the bottom 20% of the Dimensional US Large Cap Index after the exclusion of utilities, companies lacking financial data, and companies with negative relative price. A full description of the index can be found on DFA's website.

- Dimensional US Small Cap Value Index

    The Dimensional US Small Cap Value Index is compiled by Dimensional from CRSP and Compustat data. This index targets securities of US companies traded on the NYSE, NYSE MKT (formerly AMEX), and Nasdaq Global Market whose relative price is in the bottom 35% of the Dimensional US Small Cap Index after the exclusion of utilities, companies lacking financial data, and companies with negative relative price.

    The index emphasizes securities with higher profitability, lower relative price, and lower market capitalization. Profitability is measured as operating income before depreciation and amortization minus interest expense scaled by book. Exclusions: non-US companies, REITs, UITs, and investment companies. The index has been retroactively calculated by Dimensional and did not exist prior to March 2007. The calculation methodology for the Dimensional US Small Cap Value Index was amended in January 2014 to include direct profitability as a factor in selecting securities for inclusion in the index. Profitability is measured as operating income before depreciation and amortization minus interest expense scaled by book. Prior to January 1975: This index targets securities of US companies traded on the NYSE, NYSE MKT (formerly AMEX), and Nasdaq Global Market whose relative price is in the bottom 25% of the Dimensional US Small Cap Index after the exclusion of utilities, companies lacking financial data, and companies with negative relative price. A full description of the index can be found on DFA's website.

- Fama French US Large Value Research Index (provided by Fama/French from CRSP securities data)

    Composition: The index portfolios for July of year t to June $t+1$ include all NYSE, AMEX, and NASDAQ stocks for which we have market equity for December $t-1$ and June of t, and (positive) book-to-market equity data for fiscal year ending in $t-1$.

    Exclusions: ADRs, investment companies, tracking stocks, non-US incorporated companies, closed-end funds, certificates, shares of beneficial interests, and negative book values.

    Sources: CRSP databases returns and market capitalization: 1926–present; Compustat and hand-collected book values: 1926–present; CRSP links to Compustat and hand-collected links: 1926–present.

    Breakpoints: The size breakpoint is the market capitalization of the median NYSE firm, so the big and small categories contain the same number of eligible NYSE firms. The BtM breakpoints split the eligible NYSE firms with positive book equity into three categories: 30% of the eligible NYSE firms with positive BE are in Low (Growth), 40% are in Medium (Neutral), and 30% are in High (Value).

    Rebalancing: Annual (at the end of June) 1926–Present Fama/French and multifactor data provided by Fama/French.

- Fama French US Large Growth Research Index (provided by Fama/French from CRSP securities data)

Includes the upper-half range in market cap and the lower 30% in book-to-market of NYSE securities (plus NYSE Amex equivalents since July 1962 and Nasdaq equivalents since 1973). Relies, in part, on the CRSP 1–5 Index, described elsewhere.

- Fama French US Small Value Research Index (provided by Fama/French from CRSP securities data)

     Composition: The index portfolios for July of year t to June t+1 include all NYSE, AMEX, and NASDAQ stocks for which we have market equity for December t–1 and June of t, and (positive) book-to-market equity data for fiscal year ending in t–1.

     Exclusions: ADRs, investment companies, tracking stocks, non-US incorporated companies, closed-end funds, certificates, shares of beneficial interests, and negative book values.

     Sources: CRSP databases for returns and market capitalization: 1926–present; Compustat and hand-collected book values: 1926–present; CRSP links to Compustat and hand-collected links: 1926–present.

     Breakpoints: The size breakpoint is the market capitalization of the median NYSE firm, so the big and small categories contain the same number of eligible NYSE firms. The BtM breakpoints split the eligible NYSE firms with positive book equity into three categories: 30% of the eligible NYSE firms with positive BE are in Low (Growth), 40% are in Medium (Neutral), and 30% are in High (Value).

     Rebalancing: Annual (at the end of June) 1926–Present Fama/French and multifactor data provided by Fama/French.

- Fama French US Small Growth Research Index (provided by Fama/French from CRSP securities data)

     This index includes the lower-half range in market cap and the lower 30% in book-to-market of NYSE securities (plus NYSE Amex equivalents since July 1962 and Nasdaq equivalents since 1973). It relies, in part, on the CRSP 6–10 Index, described elsewhere.

- Fama French International Value Index

    2008–present: Provided by Fama/French from Bloomberg securities data. Simulated strategy of MSCI EAFE + Canada countries in the lower 30% price-to-book range. 1975–2007: Provided by Fama/French from MSCI securities data.
- Fama French International Growth Index

    Provided by Fama/French from MSCI securities data 2008–present. Provided by Fama/French from Bloomberg securities data. Simulated strategy of MSCI EAFE + Canada countries in the higher 30% price-to-book range. 1975–2007: Provided by Fama/French from MSCI securities data.
- Fama French Emerging Markets Value Index

    2009–present: Provided by Fama/French from Bloomberg securities data. Simulated strategy using IFC investable universe countries. Companies in the lower 30% price-to-book range; companies weighted by float-adjusted market cap; countries weighted by country float-adjusted market cap; rebalanced monthly. 1989–2008: Provided by Fama/French from IFC securities data. IFC data provided by International Finance Corporation.
- Fama French Emerging Markets Growth Index

    2009–present: Provided by Fama/French from Bloomberg securities data. Simulated strategy using IFC investable universe countries. Companies in the higher 30% price-to-book range; companies weighted by float-adjusted market cap; countries weighted by country float-adjusted market cap; rebalanced monthly. 1989–2008: Provided by Fama/French from IFC securities data. IFC data provided by International Finance Corporation.
- Dimensional International Small Cap Index

    The Dimensional International Small Cap Index was created by Dimensional in April 2008 and is compiled by Dimensional. July 1981–December 1993: It Includes non-US developed securities in the bottom 10% of market capitalization in each eligible country. All securities are market capitalization weighted. Each country is capped at 50%. Rebalanced semiannually. January 1994–Present: Market-capitalization-weighted index of small company securities in the eligible markets excluding those with the lowest profitability and highest relative price within the small cap universe. Profitability is measured as operating income before depreciation and amortization minus interest expense scaled by book. The index monthly returns are computed as the simple average of the monthly returns of four sub-indices, each one reconstituted once a year at the end of a different quarter of the year. Prior to July 1981, the index was 50% UK and 50% Japan. The calculation methodology for the Dimensional International Small Cap Index was amended on January 1, 2014, to include profitability as a factor in selecting securities for inclusion in the index.

- Dimensional International Small Value Index

    Small cap value companies are companies whose relative price is in the bottom 35% of their country's respective constituents in the Dimensional International Small Cap Index after the exclusion of utilities and companies with either negative or missing relative price data. The index also excludes those companies with the lowest profitability within their country's small value universe. Profitability is measured as operating income before depreciation and amortization minus interest expense scaled by book. Exclusions: REITs and investment companies. The index has been retroactively calculated by Dimensional and did not exist prior to April 2008. The calculation methodology for the Dimensional International Small Cap Value Index was amended in January 2014 to include direct profitability as a factor in selecting securities for inclusion in the index. Prior to January 1994: Created by Dimensional; includes securities of MSCI EAFE countries in the top 30% of book-to-market by market capitalization conditional on the securities being in the bottom 10% of market capitalization, excluding the bottom 1%.

All securities are market-capitalization weighted. Each country is capped at 50%; rebalanced semiannually.

- Dimensional Emerging Markets Index

    The Dimensional Emerging Markets Index is compiled by Dimensional from Bloomberg securities data. It is a market-capitalization-weighted index of all securities in the eligible markets. The index has been retroactively calculated by Dimensional and did not exist prior to April 2008.

- MSCI Country Specific Indices

    MSCI single country indexes may be used to measure the performance of large and mid-cap segments of that country's market. MSCI single country indexes cover approximately 85% of the free float–adjusted market capitalization in that country.

    The index is based on the MSCI Global Investable Market Indexes (GIMI) Methodology—a comprehensive and consistent approach to index construction that allows for meaningful global views and cross-regional comparisons across all market capitalization size, sector and style segments and combinations. This methodology aims to provide exhaustive coverage of the relevant investment opportunity set. The index is reviewed quarterly—in February, May, August, and November—with the objective of reflecting change in the underlying equity markets in a timely manner, while limiting undue index turnover. During the May and November semi-annual index reviews, the index is rebalanced, and the large and mid-capitalization cutoff points are recalculated.

- 1-Month T-Bills

    Data derived from Roger G., Ibbotson and Rex A. Sinquefield, *Stocks, Bonds, Bills, and Inflation: The Past and the Future, Dow Jones* (Chicago: Ibbotson Associates, 1989; updated).

- 1-Year Fixed Income

    Represented by the Bank of America Merrill Lynch 1-Year US Treasury Notes Index. The Bank of America Merrill Lynch 1-Year US Treasury Notes Index is an unmanaged index comprised of a single issue purchased at the beginning of the month and held for a full month. At the end of the month that issue is sold and rolled into a newly selected issue. The issue selected at each month-end rebalancing is the outstanding 2-year Treasury note that matures closest to, but not beyond, 1 year from the rebalancing date. To qualify for selection, an issue must have settled on or before the month-end rebalancing date

- Short-Term Government Bonds. Represented by ICE BofA 1 Yr. US Treasury Notes for 12/1976 to 6/1977

  The index is an unmanaged index that tracks the performance of the direct sovereign debt of the US Government having a maturity of less than 1 year.

- 1–3 Year Government Bonds. (NEW). Represented by ICE BofA 1–3 Yr. US Treasury Index July 1997–Present

  The BofA Merrill Lynch 1–3 US Year Treasury Index is an unmanaged index that tracks the performance of the direct sovereign debt of the US Government having a maturity of at least 1 year and less than 3 years. It is not possible to invest directly in an unmanaged index.

- 5-Year Government Bonds

  Represented by the Morningstar 5-Year US Treasury Notes Index (formerly Ibbotson Intermediate 5-Year Treasury Notes data), derived from stocks, bonds, bills and inflation (referenced previously).

- Bloomberg Barclays US Aggregate Bond Index (formerly Barclays Aggregate Bond Index)

  The index is a broad-based benchmark that measures the investment grade, US dollar–denominated, fixed-rate taxable bond market. The index includes treasuries, governmentrelated, and corporate securities, MBS (agency fixed-rate and hybrid ARM pass-through), ABS and CMBS (agency and non-agency). For each index, Bloomberg maintains two universes of securities: the returns (backward) and the projected (forward) universes. The composition of the returns universe is rebalanced at each month-end and represents the fixed set of bonds on which index returns are calculated for the next month. The projected universe is a forward-looking projection that changes daily to reflect issues dropping out of and entering the index but is not used for return calculations. On the last business day of the month (the rebalancing date), the composition of the latest projected universe becomes the returns universe for the following month. During the month, indicative changes to securities (credit rating change, sector reclassification, amount outstanding changes, corporate actions, and ticker changes) are reflected daily in the projected and returns universe of the index. These changes may cause bonds to enter or fall out of the projected universe of the index on a daily basis, but will affect the composition of the returns universe at month-end only, when the index is next rebalanced. Intra-month cash flows from interest and principal payments contribute to monthly index returns but are not reinvested at a short-term reinvestment rate between rebalance dates. At each rebalancing, cash is effectively reinvested into the returns universe for the following month so that index results over two or more months reflect monthly compounding.

- Barclays U.S. Intermediate Government/Credit Bond Index

  The index measures the performance of US dollar–denominated US Treasuries, government-related and investment-grade US corporate securities that have a remaining maturity of greater than 1 year and less than 10 years.

- Long-Term Government Bonds

  Data derived from Roger G., Ibbotson and Rex A. Sinquefield, *Stocks, Bonds, Bills, and Inflation: The Past and the Future, Dow Jones* (Chicago: Ibbotson Associates, 1989; updated).

- Long-Term Corporate Bonds

  Data derived from Roger G., Ibbotson and Rex A. Sinquefield, *Stocks, Bonds, Bills, and Inflation: The Past and the Future, Dow Jones* (Chicago: Ibbotson Associates, 1989; updated).

- Bloomberg U.S. Credit Bond Index Intermediate AAA

  The Bloomberg Barclays U.S. Intermediate Credit Index measures the investment grade (AAA rating), US dollar–denominated, fixed-rate, taxable corporate and government-related bond markets with a maturity greater than 1 year and less than 10 years. It is composed of the U.S. Corporate Index and a non-corporate component that includes non-US agencies, sovereigns, supranationals, and local authorities constrained by maturity. The U.S. Intermediate Credit Index is a subset of the U.S. Credit Index, which feeds into the U.S. Government/Credit Index and U.S. Aggregate Index. Index history is available back to 1973.

January 1973–Present: Bloomberg Barclays U.S. Intermediate Credit AAA Index

November 2008–August 2016: Barclays U.S. Intermediate Credit AAA Index

January 1973–October 2008: Lehman Brothers U.S. Intermediate Credit AAA Index

Total Returns in USD

Maturity: 1–10 Years

- Bloomberg U.S. High Yield Bond Index Intermediate

The Bloomberg U.S. Corporate High Yield Bond Index measures the US dollar–denominated, high-yield, fixed-rate corporate bond market. Securities are classified as high yield if the middle rating of Moody's, Fitch, and S&P is Ba1/BB+/ BB+ or below.

July 1983–Present: Bloomberg Barclays U.S. Intermediate High Yield Bond Index

November 2008–August 2016: Barclays High Yield Composite Bond Index Intermediate

July 1983–October 2008: Lehman Brothers High Yield Composite Bond Index Intermediate

Total Returns in USD

Maturity: 1–10 Years

- Gold Index

The total return for gold was calculated using the annual cumulative average price as reported by kitco.com, using New York spot gold prices quoted in US dollars per ounce.

- Commodities Index

   The Bloomberg Commodity Index is a broadly diversified commodity price index distributed by Bloomberg Indexes. The index was originally launched in 1998 as the Dow Jones–AIG Commodity Index and renamed to Dow Jones–UBS Commodity Index in 2009, when UBS acquired the index from AIG. On July 1, 2014, the index was rebranded under its current name. The index tracks prices of futures contracts on physical commodities on the commodity markets and is designed to minimize concentration in any one commodity or sector. It currently has 22 commodity futures in seven sectors. No one commodity can compose less than 2% or more than 15% of the index, and no sector can represent more than 33% of the index (as of the annual weightings of the components). The weightings for each commodity included in BCOM are calculated in accordance with rules that ensure that the relative proportion of each of the underlying individual commodities reflects its global economic significance and market liquidity.

****

Index reference information when performance data is being shown using asset class names:

U.S. Large: S&P 500 Index
U.S. Large Value: Fama/French US Large Value Research Index
U.S. Small: CRSP Deciles 6–10 Index
U.S. Micro-Cap: CRSP Deciles 9–10 Index
U.S. Small Value: Fama/French US Small Value Research Index
International Large: MSCI EAFE Index
International Small: Dimensional International Small Cap Index
International Large Value: Dimensional International Large Value Index
Emerging Markets: Dimensional Emerging Markets Index/ Fama/French Emerging Markets Index

Emerging Markets Small: Dimensional Emerging Markets Small Index

Emerging Markets Value: Dimensional Emerging Markets Value Index

Tech Stocks: Nasdaq Composite Index

Commodities Index: Bloomberg Commodity Index

International Small Value: Dimensional International Small Cap Value Index

High Quality Bonds: Bloomberg U.S. Credit Bond Index Intermediate AAA

High Yield Junk Bonds: Bloomberg U.S. High Yield Bond Index Intermediate

****

Used in correlation Figures 11.3 and 11.5:

Dimensional U.S. Large Value Index

Dimensional U.S. Small Value Index

Dimensional U.S. Micro Cap Index

Index

Page numbers followed by *f* refer to figures.